The Dawning of Gauge Theory

The Dawning of Gauge Theory

LOCHLAINN O'RAIFEARTAIGH

Princeton Series in Physics

PRINCETON UNIVERSITY PRESS • PRINCETON, NEW JERSEY

Published by Princeton University Press, 41 William Street,
Princeton, New Jersey 08540
In the United Kingdom: Princeton University Press,
Chichester, West Sussex

Library of Congress Cataloging-in-Publication Data

O'Raifeartaigh, L. (Lochlainn)
The dawning of gauge theory / Lochlainn O'Raifeartaigh.
p. cm. — (Princeton series in physics)
Includes bibliographical references and index.
ISBN 0-691-02978-4 (alk. paper). — ISBN 0-691-02977-6 (pbk. : alk. paper)
1. Quantum field theory. 2. Gauge invariance. 3. Gravitation.
4. Electromagnetism. 5. Nuclear interactions. I. Title.
II. Series.
QC174.45.07 1997
530.1′435—dc20 96-43337

This book has been composed in Times Roman

Princeton University Press books are printed on acid-free paper, and meet the guidelines for permanence and durability of the Committee on Production Guidelines for Book Longevity of the Council on Library Resources

Printed in the United States of America

1 3 5 7 9 10 8 6 4 2

1 3 5 7 9 10 8 6 4 2

(Pbk.)

Contents

PART II. The Nuclear Interactions

Preface

The quest for a unified theory of the physical world has been the goal of natural philosophy from the earliest times. Some well-known successes have been the unification of terrestrial and celestial gravity, the unification of electricity and magnetism, and the discovery that gravity was nothing but geometrical curvature. A full unification of gravitation, electromagnetism, and the other two known fundamental forces, namely, the weak and strong nuclear forces, has still to be achieved. But some steps in that direction have already been taken and the most fundamental of these has been the discovery that all of the four fundamental interactions are governed by a single principle, namely, the gauge principle.

The theory that embodies the gauge principle is known as gauge theory. One of the important features of this theory is that its interactions are mediated by mesons of the kind predicted by Yukawa in 1936: gravitation by the spin-2 graviton and the other interactions by vector mesons, namely, the photon, the gluons, and the W and Z mesons for the electromagnetic, strong, and weak interactions, respectively. An even more striking feature of gauge theory is the fact that, up to some constant parameters, the gauge principle determines the interaction of these mesons with themselves and with non-radiative matter.

The discovery of the gauge principle as a fundamental principle of physics was a slow and tortuous process that took more than sixty years. It may be convenient to separate the discovery into three separate stages.

In the first stage it was shown, mainly by Hermann Weyl, that the traditional gauge invariance of electromagnetism was related to the coordinate invariance of gravitational theory and that both were related to the gauge invariance of differential geometry. Weyl was also the first to propose that gauge invariance be elevated from the rank of a symmetry to that of a fundamental principle.

The second stage consisted in generalizing the gauge invariance used in electromagnetism to a form that could be used for the nuclear interactions. This stage began with Weyl's work and culminated in the theory that is now known as Yang-Mills gauge theory.

The third stage consisted of the gradual realization of the fact that, contrary to first appearances, the Yang-Mills gauge theory, in a suitably modified form, was suitable for describing both of the nuclear interactions.

None of the stages in the development of gauge theory were easy. Weyl's first attempt at combining electromagnetism and gravitation actually ended in disaster and was rescued only by the advent of quantum mechanics, which permitted a reinterpretation of his theory that was in accord with the experimental facts.

The second state was equally difficult. All the experimental evidence (short range of the nuclear forces, absence of neutral currents, the form of the nuclear interactions at low energy, etc.) suggested that the nuclear interactions were not gauge interactions, and the fact that any putative gauge fields would have to be electrically charged presented formidable mathematical difficulties. What is surprising is not that the theory took so long to construct but that it was constructed at all.

The third stage, the recognition of gauge theory as a reliable theory of the nuclear interactions, was as difficult as the preceding ones, though for a different reason. At this stage the difficulty was that the nature of the nuclear interactions was masked by the low-energy phenomenology, and it required the introduction of a number of new and independent concepts (parity violation, spontaneous symmetry breakdown, color symmetry, asymptotic freedom, and so on) before their true character emerged. Even today, the evidence in the case of the strong interactions is only indirect.

The third stage in the development of gauge theory is of fairly recent vintage and is thus relatively well known. The purpose of this book is to give a short sketch of the first two, lesser-known stages. The first two stages are lesser-known because, during these stages, the growth of gauge theory was not only slow and delicate but was completely overshadowed by the more spectacular developments in gravitation and quantum theory. It was only later, when much of the pioneering work had been forgotten, that the full significance of gauge theory came to be appreciated. Also, because the work done during the first stage was written in German, it was not widely accessible. Indeed one motivation for writing this book was to make this early work on gauge theory more accessible by assembling some of the relevant articles and translating them into English. The main articles included are the seminal works of Weyl in 1918 and 1929, and these are supported by articles by written by Klein, Kaluza, Fock, Schrödinger, and London in the intervening period. The list of supporting articles is, of course, not comprehensive. The choice of articles was made to mark out the path that led from Einstein's gravitational theory to Weyl's gauge theory of 1929 and to give the flavor of the general thinking on the subject at that time.

The work of Yang and Mills, which concluded the second stage in the development of gauge theory, is well known, but what is not so well known, perhaps, is that earlier studies of Yang had led to that development, and that parallel work was carried out by a number of others, notably Klein, Pauli, Shaw, and Utiyama. Recently I have been fortunate enough to obtain copies of this parallel work plus a description of his own contribution by Utiyama, and another motivation for writing this book was to assemble these contributions in an accessible manner.

The book is divided into two parts. Part I covers the development of gauge theory from the foundation laid by Einstein's theory of gravitation to Weyl's

1929 paper. Chapter 1 gives the general historical background, and in the following four chapters, the relevant article are presented, with a short description of the contents in modern terms and a commentary.[1]

Part II treats the generalization of Weyl's gauge formulation of electromagnetism to non-abelian gauge theory and contains the articles of Yang and Mills, Klein, Pauli, Shaw, and Utiyama just mentioned. Each chapter is accompanied by a commentary, which is concerned with the background and history of the articles. An interesting feature is the difference in the motivations that were used by the various authors.

It is hoped that the book will be of value not only as a description of how modern gauge theory developed in its early days but also in showing how original ideas develop and can come to fruition in spite of initial difficulties and frustrations.

[1] The descriptions of the German articles were written before it was decided to include them in translated form. Afterwards, I decided that it was better to leave the descriptions unchanged, although this would involve a certain amount of overlap with the originals. I apologize if this makes the descriptions appear a little long and detailed, especially in the case of Weyl's 1929 paper.

Acknowledgments

I am indebted to Professor C. N. Yang for a long discussion on the subject and for first drawing my attention to the papers of Schrödinger, Fock, and London presented here, and I would like to thank Professor R. L. Mills for some very welcome comments. I am indebted to Professor Norbert Straumann for copies of his contributions to the Weyl Centenary Volume and to the German Mathematical Society Centenary Volume, on both of which I have drawn heavily, and for a copy of the Pauli-Pais letters. I should like to thank Professor Pais for kind permission to include these letters and Professor Shaw for kind permission to use the relevant part of his thesis. I should also like to thank Professor Z. Ajduk for sending me the Klein extract from the Proceedings of the Kazimierz Conference. I am very grateful to Professor Misha Saveliev and Dr. Izumi Tsutsui for obtaining and translating the relevant passages of the Utiyama books and Drs Ivo Sachs and Chris Wiesendanger for checking the translations from German of the earlier papers and suggesting some improvements. I should like to thank Mr. Nicholas Bailey for proof-reading the entire manuscript and for texing the Klein, Pauli, and Shaw contributions. Finally I should like to thank the Physics Department of the University of Notre Dame for their kind hospitality for the winter semester of 1994, during which a large part of the book was written.

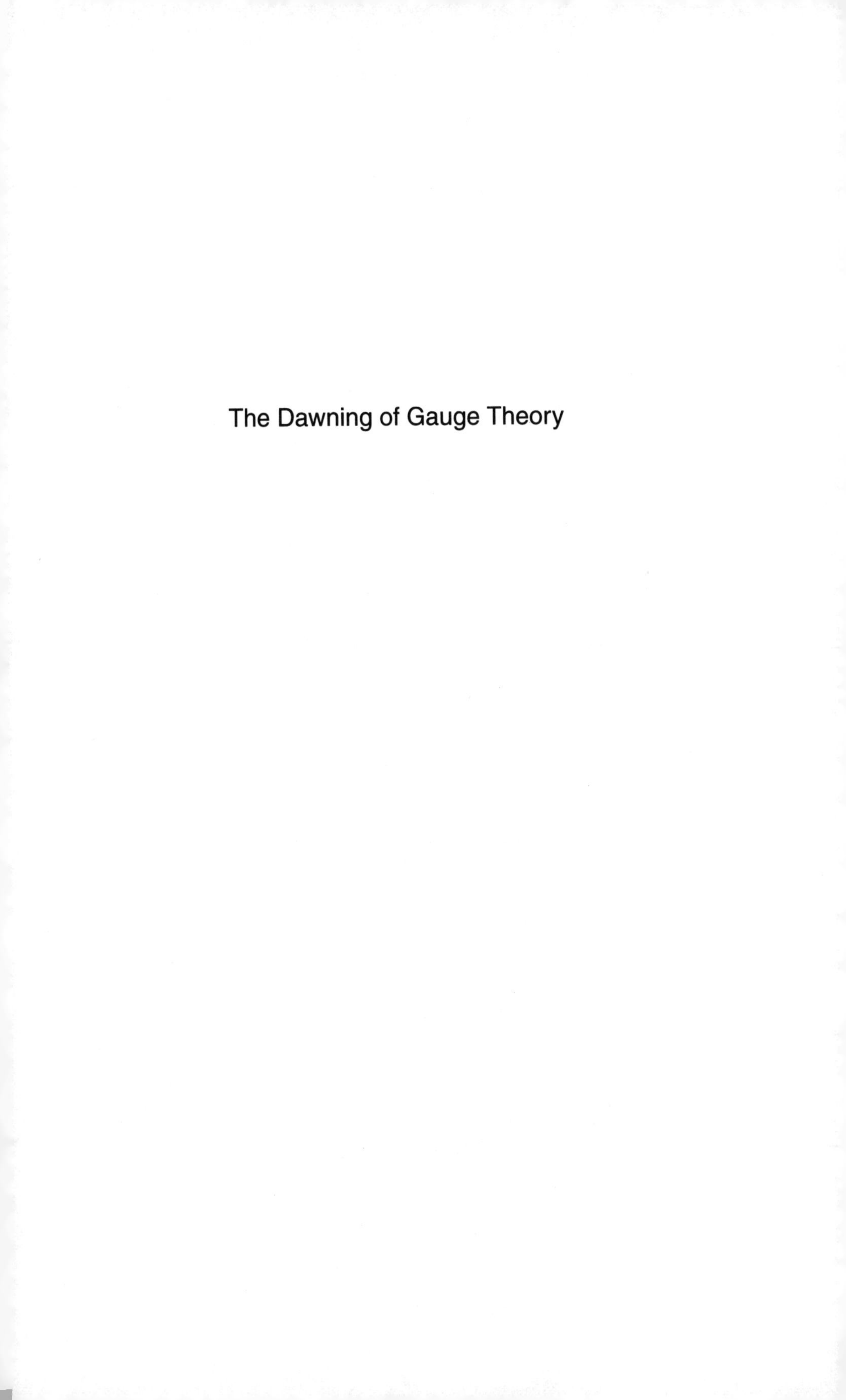

The Dawning of Gauge Theory

Introduction

The role of geometry in physics has always been central. But until the present century it was passive, providing only the stage on which the physics took place. Since the beginning of the century, however, it has come to play an ever-increasing role in the physics itself. The first entry of geometry into physics came with the theory of special relativity in 1905, when it became clear that space and time were not independent. As Minkowski [1], the inventor of space-time, put it: "From now on the separate concepts of space and time will sink into the shadows and only the union of the two will have an intrinsic meaning." But the most profound and astonishing entry of geometry into physics came with Einstein's theory of gravitation in 1916, which showed that gravity was nothing but the geometrical curvature of four-dimensional space. This idea was so simple and deep that it was appreciated even by non-scientists such as George Bernard Shaw [2], who wrote: "Asked to explain why the planets did not move in straight lines and run straight out of the universe, Einstein replied that they do not do so because space is not rectilinear but curvilinear." With these two revolutionary discoveries, due almost entirely to the genius of Einstein, geometry graduated from being the stage on which the drama of physics took place to being a major player in the drama.

There remained, however, the electromagnetic and nuclear forces, and the geometrization of gravity raised the question as to whether these other fundamental forces were "true" forces operating in the curved space of gravitational theory or whether they also were part of the geometry. This question has still not been fully answered. But what has become clear is that these forces and gravitation have a common geometrical structure. This is the so-called *gauge* structure. The purpose of this book is to explain how this structure gradually emerged.

It was actually the theory of gravitation that opened the way for the developments in physics and mathematics that led to gauge theory. Although gauge theory is now universally accepted, its geometrical nature is not always fully appreciated. This is partly because the success of gravitational theory has made the idea of geometrical forces less remarkable, partly because the geometry of gauge theory is not metrical and is therefore less intuitive, and partly because the geometry is not yet the whole story. Furthermore, the emergence of gauge theory has been a gradual process, a slow evolution rather than a revolution. The emergence of gauge theory has been gradual for two reasons. First, on the physics side, its importance for gravitation and electromagnetism was not appreciated for various reasons that will become clear later, and its role in the nuclear interactions was hidden by the phenomenology. Indeed, the short-range of the forces and the apparent absence of vector-like interactions in both nuclear forces, seemed at first to completely rule out a gauge structure. Only in the past two decades has it become clear that these were phenomenological effects due to spontaneous symmetry breaking and confinement respectively and that they masked the true situation. Second, on the mathematics side, the gauge structure that was eventually required, namely, the fiber-bundle form of differential geometry, was itself in a process of development, taking its final form only in the early fifties.

The emergence of gauge theory actually took place in two phases. In the first phase, which started with Einstein's gravitational theory and ended with Yang-Mills theory in the mid-fifties, the physicists were essentially making educated theoretical guesses, based on general principles such as symmetry, conservation laws and the need for a unified theory of gravitation and electromagnetism. What was surprising was that, with so little experimental support and so little idea about how the theory should be applied, they managed to arrive at the correct structure for the nuclear interactions. The second phase, which lasted roughly from the mid-sixties to the mid-eighties, was concerned with the application of the existing theory to the nuclear forces, whose structure had by that time become much clearer.

Each of the phases in the development of gauge theory has an interesting history, and the present book is an attempt to give a brief description of the first phase, which is perhaps less well-known [3]. The account is not meant to be comprehensive but rather to give the flavor of the early developments. Thus a large part of the book consists of original papers (most in translation) with comments. The selection is necessarily very limited and subjective, but it is hoped that it is representative.

The basic idea of gauge symmetry is that if a physical system is invariant with respect to some rigid (space-time independent) group of continuous transformations, G say, then it remains invariant when the group is made local (space-time dependent), that is, when $G \to G(x)$, where $x = x_\mu,\ \mu = 0, 1, 2, 3$ are the space-time coordinates, provided that the ordinary space-time derivatives ∂_μ are changed to covariant derivatives D_μ. The covariant derivatives D_μ take the form $D_\mu = \partial_\mu + A_\mu(x)$ where the $A_\mu(x)$ are vector fields which lie in the Lie algebra of the rigid group G and which transform so that the D_μ transform covariantly with respect to the local group. This means that invariance with respect to the local symmetry *forces the introduction of the vector fields $A_\mu(x)$ and determines the manner in which these fields interact with themselves and with matter*. The fields $A_\mu(x)$ turn out to be just the well-known radiation fields of particle physics, namely, the gravitational field, the electromagnetic field, the massive vector meson fields Z^o, $W^\pm$ of the weak interactions [4] and the coloured gluon fields A_μ^c of the strong interactions [5]. Thus gauge symmetry introduces all the physical radiation fields in a natural way and determines the form of their interactions, up to a few coupling constants.

It is remarkable that this variety of physical fields, which play such different roles at the phenomenological level, are all manifestations of the same simple principle and even more remarkable that the way in which they interact with matter is prescribed in advance. It is not surprising, therefore, to find that the covariant derivative has a deep geometrical significance. As mentioned above, modern differential geometry is formulated in terms of fiber-bundles and in this context the $G(x)$ are identified as sections of principal fiber-bundles and the radiation fields $A_\mu(x)$ as mathematical *connections*. For metrical geometry the connections are just the well-known Christoffel symbols and are secondary to the metric tensor from which they are derived, but for more general geometries the connections are the fundamental entities.

The gradual emergence of gauge theory in physics has been quite complicated. The earliest appearance was in electromagnetism, which was found empirically to be gauge-invariant. However, no geometrical significance was attached to this observation, nor was gauge invariance regarded as a fundamental principle. Indeed, it was to some extent regarded as a nuisance since its existence was inferred only indirectly from the properties of the electromagnetic field-strengths $\vec{E}(x)$ and $\vec{B}(x)$. The gauge potential $A_\mu(x)$ was tolerated because it simplified the calculations, but it had redundant components and no direct physical significance.

The next appearance of gauge invariance was in Einstein's theory of gravitation, where it appeared in the form of general-coordinate (diffeomorphic) invariance. Its appearance in Einstein's theory was to have a profound influence on the further development in two respects. First, it inspired the mathematical theory of parallel transport, which led to the theory of fiber-bundles. Second, the quest for a unified theory of gravitation and electromagnetism revealed the fact that electromagnetism had a geometrical significance and that gravitation had a gauge structure. Although the earlier attempts at a unified theory, such as Weyl's 1918 attempt to introduce electromagnetism by means of a non-integrable scale factor and Kaluza's attempt to introduce it by extending the theory of gravity to five dimensions, were unsuccessful, they paved the way for a better understanding of the role of gauge-theory. Both exhibited the geometrical structure of electromagnetism clearly and the dimensional reduction of the Kaluza theory exhibited the close relationship between coordinate and gauge transformations.

It was only after the advent of quantum theory, however, that the full significance of gauge theory emerged. It actually emerged from two quite different sources. The first source was Weyl's unsuccessful attempt to unify electromagnetism and gravity by attaching a non-integrable scale factor of the form $\exp(\frac{e}{\gamma}\int A_\mu dx^\mu)$, where γ is a constant, to the metric tensor. This proposal had been shown by Einstein to be in disagreement with observation. But after the advent of quantum theory it was suggested by London that Weyl's proposal could be used in a different manner, namely, by changing the scale factor to a *phase* factor (by choosing $\gamma = i\hbar$) and attaching it to the quantum-mechanical wave function ψ rather than the metric. Thus he proposed that the electromagnetic field be introduced by making the integral substitution $\psi \to e^{\frac{ie}{\hbar}\int A_\mu dx^\mu}\psi$. The second source was the classical minimal principle by which the momentum is changed from p_μ to $p_\mu + eA_\mu$ in the presence of an electromagnetic field. The correspondence principle of quantum theory changes p_μ to $\frac{\hbar}{i}\partial_\mu$ and thus changes the minimal principle to $\partial_\mu \to \partial_\mu + \frac{ie}{\hbar}A_\mu$. When applied to a wave equation this differential substitution is equivalent to London's integral one. The convergence of these two ideas not only had the effect of showing how the electromagnetism should be introduced at the quantum level but revealed the geometrical nature of electromagnetism because the derivative $\partial_\mu + \frac{ie}{\hbar}A_\mu$ is nothing but a covariant derivative of the kind used by mathematicians in fiber-bundle theory. The physical electromagnetic vector potential turned out to be what mathematicians call a $U(1)$ connection.

A secondary consequence of quantum theory was that the incorporation of Dirac spinors into gravitational theory required the use of the tetrad formalism in gravitation and this formalism made the gauge-structure of gravitation transparent.

The dramatic reinterpretation of his original theory introduced by these developments was fully accepted by Weyl, who then went a step further and proposed that the substitution $\partial_\mu \to \partial_\mu + \frac{ie}{\hbar}A_\mu$ be used as a *principle from which electromagnetic theory could be derived*. The principle he called the gauge principle or minimal principle. For gravity and electromagnetism the use of the gauge principle was something of a luxury since these theories were already well-established (though if it had been taken more seriously it would have led to an earlier prediction of the Aharonov-Bohm [6] effect). But for finding the structure of the nuclear interactions the gauge principle was to prove indispensible and it has become a cornerstone of modern physics. Weyl published his proposals in the 1929 paper that is reproduced here. This paper must now be regarded as one of the seminal papers of the century. Not only did it contain the gauge principle but it contained many other ideas which were innovative at the time but have since become part of the theoretical physics culture. The developments just described are summarized in Part I of the book.

Meanwhile the mathematical theory of differential geometry was progressing swiftly through the series of developments associated with names such as Levi-Civita, Cartan, de Rham, Whitney, Hodge, Chern, Steenrod and Ehresmann and ending with the construction of fiber-bundle theory [7] in the early fifties.

In spite of the importance and depth of Weyl's 1929 paper this swift mathematical development was not matched by a corresponding development in physics. The reason was simple: Weyl had already exhausted the possibilities for gravity and electromagnetism and the nuclear interactions showed no sign of having anything to do with gauge theory. On the contrary, the experimental evidence suggested that they were mediated by scalar and fermionic interactions. Furthermore, if gravitation and electromagnetism were any guide, the gauge fields would have to be massless and the short range of the nuclear forces completely ruled this out. It was only when, in the wake of parity-violation, the $V - A$ structure of the weak interactions emerged in 1958, that the vectorial structure of the weak interactions become apparent, and it was only after the discovery of the asymptotic freedom of non-abelian gauge theories in 1974 that the possibility that the strong interactions might be vectorial was taken seriously. In the period 1929-56 the gauge-theoretical picture of the world seemed to be ruled out by the phenomenology. The authors who finally constructed the theory were swimming against the phenomenological tide.

In view of this it is not surprising that the developments in physics were much slower [8] than in mathematics and that, if one discounts the somewhat fortuitous construction of an $SU(2)$ gauge theory in 1938 that will be discussed later, no attempt was made to construct a gauge theory for the nuclear interactions until the early fifties, when Yang and Mills constructed the $SU(2)$ gauge theory which has become synomymous with their names.

The Yang-Mills theory is well-known, but what is, perhaps, not so well-known is that it originated in ideas that went back to Yang's graduate student days in Chicago and that there were parallel constructions by other physicists. All these constructions, especially that of Yang-Mills, stemmed from Weyl's 1929 paper and were completely independent of the mathematical developments just mentioned. Only later did the relationship between the physics and the geometry become manifest.

The idea of Yang just mentioned was that the isospin symmetry of the strong interactions should be gauged in the same way as the electromagnetic phase symmetry, an idea that was inspired by reading of Weyl's ideas in Pauli's *Handbuch* article on quantum theory. As discussed later, Yang made some immediate progress on the problem, but was held up for a time by his failure to find the correct expression for the field strengths, a problem that was solved in collaboration with Mills in the summer of 1953. Yang and Mills did not immediately publish their results because they were aware of the difficult problems posed by the gauge-field masses and renormalization. After studying these problems for some time and realizing that they would not be solved in the short term, they sent their paper for publication in the spring of 1954.

As mentioned earlier, some similar work was carried out by other physicists, who had quite different motivations. Although the other contributions were not quite as early, or as decisive, as the Yang-Mills result, they came within months of it and produced the same results quite independently. They are good examples of how the same result can be obtained using completely different starting points and different methods. For this reason, and because they also illustrate how original ideas develop, I have thought it worthwhile to include these other contributions here.

One such contribution actually predated the Yang-Mills theory by about fifteen years [9] and came in a paper by Klein at a conference in Kasimierz, Poland, in 1938. Klein was stimulated by Yukawa's suggestion of two years earlier that the nuclear interactions might proceed through the exchange of mesons. Aware that at least some of the Yukawa particles would have to be electrically charged to account for the charge exchange of the strong nuclear interactions and assuming that they would be vector fields like the photon, Klein proposed a Kaluza-type model in which the extra components of the metric tensor were regarded as components of a 2×2 matrix, similar to the isotopic spin matrix of nuclear physics. In the best sleepwalker [10] tradition he thereby stumbled onto what we would now call an $SU(2)$ gauge theory of the strong interactions. The combination of electromagnetic gauge invariance, the Kaluza reduction and the 2×2 matrix character of the metric were enough to produce $SU(2)$ gauge invariance. What was astonishing was that Klein used only the electromagnetic gauge principle and even after constructing the theory was apparently unaware of its $SU(2)$ gauge invariance. Indeed, he quite happily violated the $SU(2)$ gauge invariance by giving masses to the charged vector fields. He even extended his theory to what we would now call an $SU(2) \times U(1)$ gauge theory in order to take the charge independence of the strong interactions into account. Because it was not published in a regular journal and because of the outbreak of the Second World War, Klein's paper went largely unnoticed, and it was not until 1953 that the links between gauge theory and the nuclear interactions resurfaced.

The other three contributions which are included were almost contemporaneous with Yang and Mills. Indeed, they were so close that, although there is no doubt about Yang-Mills's priority, it was achieved by only a short head.

The first of the other contributors was Pauli, who, inspired by a lecture on the strong interactions given by Pais at another conference, drafted a paper called *Meson-Nucleon Interactions from a Differential Geometric Point of View* in which he arrived at the essentials of an $SU(2)$ gauge-theory by using dimensional reduction. Pauli did not appear to be convinced by his results, mainly, it would seem, because of the fermionic spectrum and he did not publish them, or even have them typed. But he sent handwritten versions to Pais in July and December 1953, copies of which are included here. A particularly interesting feature of Pauli's construction was that the dimensional reduction automatically produced the expressions for the field strengths that had given Yang such trouble. It is clear from the letters that Pauli regarded the derivation of these expressions as his main result.

Meanwhile, at Cambridge University, Ronald Shaw, who is now professor of mathematics at the University of Hull but was then a doctoral student of Abdus Salam at Cambridge, was deriving the theory in yet another way. Shaw was inspired by a preprint of Schwinger's in which the electromagnetic gauge group $U(1)$ was presented in its 2-dimensional real $SO(2)$ form. He says that on reading this preprint he "realized in a flash" that the theory could be generalized to $SU(2)$, and he then proceeded to write down what was essentially the same theory as that of Yang and Mills. However, for various reasons that will be explained later, the results were not published until September 1955 and then only in his doctoral thesis.

Finally, at about the same time as Shaw, the Japanese physicist Utiyama was assembling the ideas on gauge theory which he had developed from the tetrad formulation of gravitation, in preparation for a visit to Princeton. Through his studies he had become aware that the common feature of gravitation and electromagnetism was the mathematical *connection*. He considered that, at some as yet unknown level, this feature might be common to all fundamental interactions, and he therefore constructed the gauge theory for all Lie groups. Unfortunately, on his arrival in Princeton in April 1954 he received a preprint of the Yang-Mills work and was so stunned by its resemblance to his own that he shelved his results for a year. It was only when he realized that Yang and Mills had considered only the $SU(2)$ group, and had not considered gravitation, that he sent his work for publication, with the result that it was not published until 1956. Because Utiyama considered general groups and quoted Yang and Mills, this 1956 paper is often thought to be a straightforward generalization of the Yang-Mills theory. But as can be seen from the extracts from Utiyama's books presented here, this is not the case.

The recounting here of the three independent constructions of non-abelian gauge theory is not meant to detract from the originality and importance of the Yang-Mills paper, which is justly celebrated as the first systematic construction and application of non-abelian gauge theory in physics. Yang and Mills had completed their theory in 1953 and published it in 1954, in contrast to Shaw and Utiyama who completed their theories in early 1954 and did not publish until late 1955 and 1956, respectively, and to Pauli, who wrote his first draft in the summer of 1953 but never had it typed, much less published. Thus Yang and Mills have priority with respect to both conception and publication. Their paper was also the only one to be published in a regular journal and they were the only ones, except for Pauli, to consider the application to the nuclear interactions seriously. But it was a close race.

This book divides naturally into the two parts already mentioned. Part I deals with the period from Einstein's gravitational theory (1916) to Weyl's seminal 1929 paper and is concerned only with electromagnetism and gravity. This period saw the gradual emergence and appreciation of the gauge structure of electromagnetism, especially within the context of quantum theory and culminated with Weyl's formulation of the minimal principle in 1929. It includes translations of original papers from the period and, as some of these papers are difficult to peruse from a modern point of view, I have attempted to summarize the salient points. An interesting feature of these papers, and of the later papers by Klein and Pauli, is the role played by the attempts to unify gravitation and electromagnetism [9], particularly the attempts based on dimensional reduction, which provided a method of formulating the abstract notion of gauge potentials in a familiar metrical manner.

Part II of the book deals with the attempts to find the non-abelian analogue of electromagnetism with a view to describing the nuclear interactions. It consists of a presentation of the papers of Yang-Mills, Klein, and Shaw, together with a typescript of the Pauli letters and some historical comments on his own work extracted from books written by Utiyama.

PART I

Gravitation and Electromagnetism

1

Gauge Transformations in Classical Electromagnetism

It is probably fair to say that gauge theory originated in the discovery that, in contrast to the Newtonian gravitational and static electrical fields, which are produced by masses and and electric charges respectively, the magnetic field is not produced by magnetic charges but by magnetic dipoles. This pecularity of the magnetic field $\vec{B}(x)$ is expressed mathematically by the statement that the magnetic field is divergence-free throughout space,

$$\vec{\partial}.\vec{B}(x) = 0 \quad \text{everywhere.} \tag{1}$$

This means that the magnetic field cannot be the gradient of a scalar field like the gravitational and electrical fields, which are of the well-known form

$$\vec{E}(x) = \vec{\partial}\phi(x), \tag{2}$$

but is the curl of a vector field,

$$\vec{B}(x) = \vec{\partial} \wedge \vec{A}(x). \tag{3}$$

The curl property of the magnetic field was known quite early and led, for example, to Stokes' theorem

$$\int \vec{B}.d\vec{S} = \oint \vec{A}.d\vec{s}, \tag{4}$$

where S is a two-dimensional surface with a closed boundary ∂S and the line-integral on the right-hand side is taken around ∂S. We shall see that the Stokes line-integral played an important role in the history of gauge theory. Indeed, it continues to play a central role in that theory, appearing, for example, in the Aharonov-Bohm phase factor in electromagnetism and in the Wu-Yang phase factor [1] and the Wilson loop [2] in non-abelian gauge-theory.

There are a number of new features associated with the properties (1) and (3) of the magnetic field. First, since the differentiation in (3) is outer it is independent of the metric and is valid even if the space is curved. The only restriction is that if the space is not topologicaly trivial then (3) may be true only locally. Second, the economy that is gained by expressing the gravitational and electric fields as the gradients of a single scalar field $\phi(x)$ is lost because $\vec{A}(x)$ has as many components as $\vec{B}(x)$. Third, and perhaps most importantly, $\vec{A}(x)$ is unique only up to the addition of the gradient of a scalar *function*, in contrast to the scalar potential $\phi(x)$, which is unique up to a *constant*. Thus the magnetic field $\vec{B}(x)$ field remains invariant under the transformations

$$\vec{A}(x) \quad \rightarrow \quad \vec{A}(x) + \vec{\partial}\alpha(x), \tag{5}$$

where the $\alpha(x)$ are any differentiable scalar functions.

Because of the analogy between the roles of $\vec{A}(x)$ in magnetism and $\phi(x)$ in electricity the field $\vec{A}(x)$ is often called the vector potential. But for historical reasons that will become clear later it is also called the *gauge* potential, and the transformations (5) are called *gauge* transformations. It is remarkable that these transformations of a field whose existence is only inferred mathematically from the properties of the observed magnetic field $\vec{B}(x)$ have come to play the central role in the theory of the fundamental forces.

Since the sequence of two transformations (5) with any differentiable functions $\alpha_1(x)$ and $\alpha_2(x)$ is equivalent to a single transformation of the same kind with the differentiable function $\alpha(x) = \alpha_1(x) + \alpha_2(x)$ we see that for any fixed x the transformations (5) form a group, and that it is a commutative (abelian) group with a single continuous parameter $\alpha(x)$. It is customary to call this group the gauge group. The topology of the gauge group (whether the parameter $\alpha(x)$ takes its values on the real line or the circle) will be discussed later.

RELATIVISTIC GENERALIZATION

As is well-known, the above considerations generalize in a simple and natural way to the relativistic case. Indeed, in that case the divergence-free property that was originally associated with the magnetic field generalizes to a property of the electromagnetic field as a whole. To see this, let μ, ν and i, j denote the usual 4-dimensional and 3-dimensional indices respectively, and let the electric and magnetic fields $\vec{E}$ and $\vec{B}$ be written as a single anti-symmetric Lorentz tensor $F_{\mu\nu}$, where $F_{oi} = E_i$ and $F_{ij} = \epsilon_{ijk}B_k$. Then equation (1) generalizes to

$$\partial \wedge F = 0 \quad \text{or} \quad \epsilon^{\mu\nu\lambda\sigma}\partial_\nu F_{\lambda\sigma} = 0, \tag{6}$$

where the indices run from 0 to 3 and ϵ is the 4-dimensional numerical Levi-Civita symbol. Furthermore, just as the divergence-free character of the magnetic field implies that it is locally the curl of a 3-vector, the condition (6) implies that the electromagnetic field is locally the curl of a 4-vector

$$F = \partial \wedge A \quad \text{or} \quad F_{\mu\nu} = \partial_\mu A_\nu - \partial_\nu A_\mu \quad \text{where} \quad A_\mu = \{\vec{A}(x), \phi(x)\}. \quad (7)$$

Thus the 4-vector A_μ is just a combination of the 3-vector and scalar potentials.

As in the 3-dimensional case the derivative in (6) is outer and thus it is locally valid in curved spaces and globally valid if they are topologically trivial. Note that in 4 dimensions the derivatives in both (6) and (7) are outer, and thus both equations are independent of the metric. Also, in four dimensions some economy is regained by using the vector potential because A_μ has only four components whereas $F_{\mu\nu}$ has six. More generally, in d dimensions the use of the vector potential reduces the number of components from $\frac{1}{2}d(d-1)$ to d. Finally, just as in the 3-dimensional case, the field A_μ is defined only up to the gradient of a scalar field. That is, $F_{\mu\nu}$ remains unchanged when A_μ undergoes a transformation of the form

$$A_\mu(x) \quad \rightarrow \quad A_\mu(x) + \partial_\mu \alpha(x), \quad (8)$$

for any differentiable Lorentz scalar $\alpha(x)$. The transformations (8) are clearly the 4-dimensional generalizations of the gauge transformations (5) and have the same kind of group properties.

LAGRANGIAN AND HAMILTONIAN

The ambiguity in the definition (7) of $A_\mu(x)$ is not removed by the Maxwell field equations

$$\partial_\mu F^{\mu\nu} = j^\nu, \quad (9)$$

where j^ν is the electric-current, or by the Lorentz equations for the motion of a particle in an electromagnetic field,

$$\frac{dp^\mu}{d\tau} = F^\mu_{\ \nu} \frac{dx^\nu}{d\tau}, \quad (10)$$

where τ is the proper time, since both of these use only the electromagnetic field strengths $F_{\mu\nu}$. Accordingly, in the classical theory at least, the vector potential $A_\mu(x)$ has no direct physical meaning.

Since the reduction from six to four in the number of fields is offset by the ambiguity in the definition of $A_\mu(x)$, one might then ask why one bothers to introduce $A_\mu(x)$ in the first place. There are a number of reasons for doing so.

(i) The field $A_\mu(x)$ is unconstrained whereas the field $F_{\mu\nu}$ is constrained by the condition (6). Hence the use of $A_\mu(x)$ provides a means of satisfying the constraint identically.

(ii) The electromagnetic (Maxwell) field equations can be obtained from a variational principle only if the Lagrangian

$$L = \frac{1}{4}\int F^{\mu\nu}F_{\mu\nu} = \frac{1}{2}\int\left(\vec{E}^2 - \vec{B}^2\right), \tag{11}$$

is varied with respect to A_μ. Variation with respect to the $F_{\mu\nu}$ themselves would result in trivial algebraic equations.

(iii) The Lagrangian and Hamiltonian for a particle in an electromagnetic field

$$\mathcal{L} = m\sqrt{\left(\frac{dx^\mu}{d\tau}\frac{dx_\mu}{d\tau}\right)} + eA_\mu\frac{dx^\mu}{d\tau} \quad \text{and} \quad \mathcal{H} = \sqrt{(\vec{\pi} - e\vec{A})^2 + m^2} + e\phi(x), \tag{12}$$

respectively, where

$$\vec{\pi} \equiv \frac{\partial L(t)}{\partial\frac{d\vec{x}}{d\tau}} = \vec{p} + e\vec{A}(x) \qquad \vec{p} = m\frac{\frac{d\vec{x}}{d\tau}}{\sqrt{\left(\frac{dx^\mu}{d\tau}\frac{dx_\mu}{d\tau}\right)}}, \tag{13}$$

from which the Lorentz-force equations are derived, are *local* functions of $A_\mu(x)$ but cannot be expressed as local functions of $F_{\mu\nu}(x)$. Thus the use of the vector potential $A_\mu(x)$ is natural for the Lagrangian and Hamiltonian formulation of electromagnetic theory.

The reason that the Lagrangian, which contains $A_\mu(x)$, can produce a gauge-invariant Lorentz force equation is that the Lagrangian is invariant with respect to the gauge transformations (8), in the sense that under such a transformation it changes only by the total derivative

$$e\frac{\partial\alpha(x)}{\partial x_\mu}\frac{dx^\mu}{d\tau} = e\frac{d\alpha}{d\tau}, \tag{14}$$

and this leaves the variational equations unchanged. The Hamiltonian formulation is gauge-invariant because $\mathcal{H}$ is actually a function of $\vec{p}$, and, as can be seen from the Lorentz force equation (10), $\vec{p}$ is gauge-invariant.

THE MINIMAL PRINCIPLE

From the Hamiltonian formulation of the motion of a charged particle in an electromagnetic field and the fact that $H = -p_o$, one sees that the correct

Lorentz-force equations are derived by taking the free Hamiltonian and making the substitution

$$p_\mu \quad \rightarrow \quad p_\mu + eA_\mu(x). \qquad (15)$$

The principle by which the effect of an electromagnetic field on a particle of charge e is obtained by changing p_μ to $p_\mu + eA_\mu(x)$ is called the *minimal principle*. The word *minimal* refers to the fact that although the gauge principle implies gauge invariance, it is stronger than gauge invariance alone since it excludes direct interactions with the gauge invariant electromagnetic field strengths $F_{\mu\nu}$. Present experimental evidence shows that the use of the minimal principle is justified for fundamental fields such as the leptons and quarks but not for composite fields such as the nucleons. For the latter the electromagnetic form factors show that there are direct interactions with the field strengths $F_{\mu\nu}$ as well as a minimal interaction with the vector potential.

The minimal principle encapsulates in an economical way the information contained in the Lorentz field equation and thus provides yet another reason for using the gauge potential. But its use goes beyond mere economy. As might be expected from the fact that it applies to fundamental fields, it is actually one of the fundamental principles on which gauge theory is based. It played a crucial role in the development of abelian and non-abelian gauge theory, as we shall see, and it continues to play a central role in these theories.

FEYNMAN'S DERIVATION OF ELECTROMAGNETISM

Although electromagnetic theory in its final form is quite elegant and is encapsulated by the minimal principle, its discovery, apart perhaps from Maxwell's prediction of the displacement current, was empirical. This prompts the question as to whether electromagnetism, like gravitation, could be derived from some more universal principle. No such principle is known at present but it might be interesting to mention some earlier work of Feynman that has recently been published and throws some light on the question.

Feynman was actually looking for some new type of fundamental force and for this purpose decided to investigate what would happen if the canonical commutation relations (or Poisson brackets) for the momentum variables $\vec{p}$ were relaxed. To his surprise and disappointment, he found that the force obtained in this way is just the conventional electromagnetic force and did not publish his results. But, for historical reasons, and because the result could also be interpreted positively as a derivation of electrodynamics, the Feynman argument was recently published by Dyson [3]. It may be stated briefly as follows:

The commutation relations which are not relaxed are

$$[x_a, x_b] = 0 \qquad [p_a, x_b] = i\delta_{ab} \quad \text{where} \quad a, b = 1, 2, 3, \tag{16}$$

and the first question is what commutation relations for the momenta $\vec{p}$ are permitted by these. From the Jacobi relations one finds without much difficulty that the most general commutation relations permitted are

$$[p_a, p_b] = -\epsilon_{abc} B_c(x) \quad \text{where} \quad \partial_c B_c(x) = 0. \tag{17}$$

Since (16) and (17) together constitute a symplectic structure there are no further conditions. The last equation in (17) implies the existence of a vector potential $\vec{A}$ and if one defines $\vec{\pi} = \vec{p} + \vec{A}$ one finds that the π_i commute. Thus (16) and (17) may also be written in the form

$$[x_a, x_b] = 0 \qquad [p_a, x_b] = i\delta_{ab} \qquad [\pi_a, \pi_b] = 0 \quad \text{where} \quad \vec{p} = \vec{\pi} - \vec{A}(x). \tag{18}$$

These are just the canonical commutation relations supplemented by the minimal principle $\vec{\pi} = \vec{p} + \vec{A}(x)$ (e normalized to unity). Thus the relaxation of the canonical commutation relations provides just enough freedom for the introduction of the 3-potential $\vec{A}(x)$.

To include the electric field one needs to consider the relationship between p_a and $\dot{x}_a$, which is equivalent to choosing the kinetic part of the Hamiltonian. The relationship actually chosen by Feynman was the usual non-relativistic one $p_a = m\dot{x}_a$ (indeed he wrote the commutation relations in terms of $m\dot{x}_a$ rather than p_a) and it is easy to see that this choice leads to the conventional non-relativistic electromagnetic Hamiltonian

$$H = \frac{1}{2m}(\vec{p})^2 + \phi(x) = \frac{1}{2m}\left(\vec{\pi} - \vec{A}(x)\right)^2 + \phi(x), \tag{19}$$

where the potential $\phi(x)$ is completely free. (The relativistic analogue (19) is obtained by replacing m by its relativistic counterpart $\sqrt{m^2 + (\vec{p})^2}$ in the relationship between p_a and $\dot{x}_a$).

Thus from Feynman's investigation one sees that conventional electromagnetism and the minimal principle may be regarded as expressions of the freedom that is obtained when canonical commutation relations for the momenta and the potential part of the Hamiltonian are left open. Whether this freedom or the minimal principle itself should be regarded as the more fundamental is, of course, open to debate. In any case Feynman's result indicates how difficult it is to find a plausible alternative to gauge theory.

HAMILTON-JACOBI EQUATION

We conclude this discussion by recalling for future reference two concepts which played a central role in the historical development of gauge theory, namely, the Hamilton-Jacobi equation and the Noether conservation theorem.

In this section we consider the Hamilton-Jacobi equation, which in standard non-relativistic Hamiltonian theory takes the form

$$\dot{W} + \frac{1}{2m}(\vec{\partial} W)^2 + V = 0 \quad \text{where} \quad \vec{\partial} W = \vec{p} \quad \text{and} \quad \dot{W} = -H. \tag{20}$$

The importance of the Hamilton-Jacobi function W is that the Hamiltonian equation of motion

$$\partial_t \vec{p} = -\frac{\partial H}{\partial \vec{x}}, \tag{21}$$

may be regarded as the necessary and sufficient conditions for its existence. In other words, the Hamiltonian equation (21) may be regarded as the integrability condition for the last two equations in (20).

For non-relativistic particles moving in an electromagnetic field the Hamilton-Jacobi equation becomes

$$\left(\dot{W} - e\phi(x)\right) + \frac{1}{2m}\left(\vec{\partial} W - e\vec{A}\right)^2 + V(x) = 0, \tag{22}$$

and, as might be expected, the relativistic generalization of (22) is

$$\left(\partial_\mu W - eA_\mu(x)\right)^2 + (m+V)^2 = 0 \qquad \text{where} \qquad \partial_\mu W - eA_\mu = \frac{dx_\mu}{d\tau} = p_\mu. \tag{23}$$

Note that the electromagnetic Hamilton-Jacobi equation is obtained from the neutral one by making the substitution

$$\partial_\mu W \quad \rightarrow \quad \partial_\mu W - eA_\mu(x), \tag{24}$$

which is therefore the Hamilton-Jacobi form of the minimal principle. This implies, of course, that the Hamilton-Jacobi function is not gauge-invariant but gauge-transforms according to

$$W(\vec{x}, t) \qquad \rightarrow \qquad W(\vec{x}, t) + e\alpha(x). \tag{25}$$

We shall see later that the Hamilton-Jacobi form of the minimal principle is an important foreshadowing of the way in which it appears in quantum theory. Indeed the Hamilton-Jacobi form of the principle was used to introduce electromagnetism into quantum-mechanics.

For later use we note that, with suitable initial conditions, the Hamilton-Jacobi equation can be integrated to express W in the form

$$W = m\tau + e\int_o^x A_\mu(y)dy^\mu, \tag{26}$$

where the integral is taken along the path defined by $\frac{dy^\mu}{d\tau}$ in (23). To see this we note that

$$\begin{aligned} W - e\int_o^x A_\mu(y)dy^\mu &= \int_o^x (\partial_\mu W - eA_\mu(y))dy^\mu \\ &= \int_o^x p_\mu dy^\mu = \frac{1}{m}\int_o^\tau p_\mu p^\mu d\tau = m\tau, \end{aligned} \tag{27}$$

where the last equation follows from the Hamilton-Jacobi equation.

NOETHER THEOREM

The Noether theorem [4] gives the general relationship between symmetries and conservation laws. In its simplest form it is simply the observation that the commutator or Poisson-bracket relationship $[H, Q] = 0$ between the Hamiltonian and any dynamical variable Q has a dual interpretation: considered as the action of H on Q it implies that Q is conserved and considered as the action of Q on H it implies that Q generates a symmetry group. Thus to every symmetry there corresponds a conserved quantity and conversely.

Noether's theorem formulates this result for continuous groups in the context of Lagrangian field theory and conserved currents. Let $L(\phi(x), \partial\phi(x))$ be a Lagrangian density for fields $\phi(x)$ of any spin and consider any infinitesimal variation $\delta\phi(x)$ of the fields which does not involve a change in the coordinates.

Letting π^μ denote $\frac{\partial L}{\partial\phi_{,\mu}}$ the variation of L is then

$$\delta L = \pi^\mu\partial_\mu\delta\phi + \frac{\partial L}{\partial\phi}\delta\phi = \partial_\mu(\pi^\mu\delta\phi) + \frac{\delta L}{\delta\phi}\delta\phi, \tag{28}$$

where $\frac{\delta L}{\delta\phi}$ is the variation that produces the usual field equations. Indeed the field equations are produced by the demand that δL vanish when the variation of ϕ is zero at the boundary. Hence when the field equations are satisfied and the variation $\delta\phi(x)$ does not vanish at the boundary we have

$$\delta L = \partial_\mu j^\mu \quad \text{where} \quad j^\mu = \pi^\mu\delta\phi \qquad \text{(on-shell)}, \tag{29}$$

where "on-shell" means that the field equations are satisfied. Equation (29) is an on-shell identity that holds for any variation. But it shows that if the variation corresponds to a symmetry of the Lagrangian density, $\delta L = 0$, the current j^μ is conserved. In other words,

$$\delta L = 0 \quad \Leftrightarrow \quad \partial_\mu j^\mu = 0 \quad \text{for} \quad j^\mu = \pi^\mu \delta\phi \quad \text{on shell.} \tag{30}$$

This is Noether's theorem in its basic form. The relationship of (30) to conserved charges Q is that for suitable boundary conditions

$$\partial_\mu j^\mu = 0 \quad \Rightarrow \quad \dot{Q} = 0 \quad \text{where} \quad Q = \int d^3x j^o(x). \tag{31}$$

A corollary to the theorem is that the Poisson-bracket or commutator action of the charges Q on the fields will produce their infinitesimal variations. Thus the conserved charges implement the symmetry.

The application of Noether's theorem to gauge and Poincare symmetries is not quite straightforward, and since they are important for later discussions it may be worthwhile to consider them in a little more detail.

First, for gauge groups it is actually only the rigid subgroups, i.e., the subgroups with space-time independent parameters, that contribute to Noether's theorem. The point is that if we consider the subgroup $\tilde{G}(x)$ of the gauge group such that $\tilde{G}(x) \to 1$ as $x \to \infty$ the variations of the fields due to $\tilde{G}(x)$ vanish at infinity and are therefore a subset of the variations that yield the field equations. Hence they cannot give an on-shell contribution to (29). Since every element of the gauge group can be decomposed into the product of an element of $\tilde{G}(x)$ and an element of the rigid subgroup G it follows that the contributions to (29) can come only from the rigid subgroup. In fact, each rigid one-parameter subgroup yields a non-zero conserved current and thus for an n-parameter gauge group there are just n Noether conserved charges.

For the Poincare group we consider translations and Lorentz transformations seperately. For translations the variation of the Lagrange density is no longer zero but a divergence of the form $\delta L = \epsilon^\alpha \partial_\alpha L$ where the ϵ's are the four (rigid) translation parameters. Hence (30) has to be modified to

$$\partial_\mu T^{\mu\alpha} = 0 \quad \text{where} \quad T^{\mu\alpha} = j^{\mu\alpha} - g^{\mu\alpha} L. \tag{32}$$

The current $T^{\mu\alpha}$ is, of course, the energy momentum tensor density and the conserved charges associated with it are the total energy and linear momentum.

For Lorentz transformations the Lagrangian density L is invariant but the variation of the derivatives ∂_μ in $L(\phi, \partial_\mu \phi)$ must be taken into account. Hence (30) is no longer valid but becomes

$$\partial_\mu \left(\omega_{\alpha\beta} j^{\mu\alpha\beta} \right) = \omega_{\alpha\beta} T^{\alpha\beta}, \tag{33}$$

where we combine the individual Lorentz Noether currents in a single current $j^\mu = j^{\mu\alpha\beta}\omega_{\alpha\beta}$ where $\omega_{\alpha\beta}$ is the anti-symmetric matrix of the (rigid) Lorentz parameters and $T^{\alpha\beta}$ is the energy momentum tensor density. Thus only if the energy momentum tensor is symmetric is there a conserved current (in which case the conserved charges are the angular momentum and its relativistic counterpart). For scalar fields $T^{\alpha\beta}$ is automatically symmetric but for higher spin fields this may not be the case; for these fields T has to be redefined so that it is symmetric or modified by the addition of a divergence to make it symmetric [5].

The above considerations are classical but, as might be expected, there are generalizations for quantized fields. The well-known Ward-Takahashi and Taylor-Slavnov [6] identities of abelian and non-abelian gauge theories are the most common generalizations. Furthermore the cases in which the Noether current-conservation breaks down in quantum theory are of great interest. The breakdown occurs because the quantum theory is defined not only by the Lagrangian, but by the measure in the quantum-mechanical functional integral and by the renormalization procedure. For a quantum-mechanical symmetry to hold all three of these must be invariant. The cases in which the Lagrangian is invariant but the measure is not lead to the so-called *anomalies* [7], which play a central role in modern quantum field theory.

HISTORICAL ROLE OF GAUGE INVARIANCE

Although the gauge-invariant formulation of electrodynamics was quite standard by the beginning of this century, the true significance of gauge invariance does not seem to have been appreciated. Indeed, its geometrical significance was not yet understood, and there was certainly no thought of using gauge invariance or the minimal principle as fundamental principles. For example, the modified version of electrodynamics proposed by Mie [8] in 1912 was criticized on the grounds that it did not agree with experiment but the disagreement was not traced to the fact that it was not gauge-invariant.

That even in the mid-twenties gauge invariance was by no means taken for granted is shown by the fact that Schrödinger in his famous 1926 papers made no explicit use of gauge invariance, and that Fock as late as 1927 was quite surprised and gratified to find that the gauge transformation (5) left the Balmer spectrum invariant (see the work of these authors reproduced here).

It is also, perhaps, interesting to find that even Einstein, in his search for the gravitational field equations, lost some time in 1913 by failing to take gauge invariance (in this case coordinate invariance) into account. He thought at the time that the field equations $G_{\mu\nu} = \kappa T_{\mu\nu}$, where $T_{\mu\nu}$ is the energy-momentum tensor, should determine the metric uniquely and, on finding that they could not do so (because the energy-momentum conservation laws reduce the number of independent components of $T_{\mu\nu}$ to six) he came to the conclusion [9] that the law of general covariance should be restricted. What he had overlooked was the fact that the components of the metric tensor should be determined only up to gauge (coordinate) transformations and the energy-momentum conservation laws were just an expression of this fact. Einstein soon discovered his error, of course, and proceeded to his celebrated theory of gravitation three years later.

Gravitation and Electricity

by H. Weyl in Zurich

Sitzungsber. Preuss. Akad. Berlin (1918) 465

According to Riemann [1] geometry is based on the following two facts:

1. *Space is a three-dimensional continuum*, the manifold of its points is therefore represented in a smooth manner by the values of three coordinates x_1, x_2, x_3.

2. (*Pythagorean Theorem*) The square of the distance between two infinitesimally separated points

$$P = (x_1, x_2, x_3) \qquad \text{and} \qquad P' = (x_1 + dx_1, x_2 + dx_2, x_3 + dx_3) \tag{1}$$

is (in any coordinate system) a quadratic form in the relative coordinates dx_i:

$$ds^2 = \sum_{ik} g_{ik} dx_i dx_k \qquad (g_{ik} = g_{ki}) \tag{2}$$

We express the second fact briefly by saying: the space is a *metrical* continuum. In the spirit of modern local physics we take the pythagorean theorem to be strictly valid only in the infinitesimal limit.

Special relativity leads to the insight that *time* should be included as a fourth coordinate x_o on the same footing as the three space-coordinates, and thus the stage for physical events, *the world*, is a *four-dimensional, metrical continuum*. The quadratic form (2) that defines the world-geometry is not positive-definite as in the case of three-dimensional geometry, but has positive-index 3. Riemann already expressed the idea that the metric should be regarded as something physically meaningful since it manifests itself as an effective force for material bodies, in centrifugal forces for example, and that one should therefore take into account that it interacts with matter; whereas previously all geometers and philosophers believed that the metric was an intrinsic property of the space, independent of the matter contained in it. It was on the basis of this idea, for which the possibility of fulfillment was not available to Riemann, that in our time Einstein (independently of Riemann) erected the grandiose structure of general relativity. According to Einstein the phenomena of gravitation can be attributed to the world-metric, and the laws through by which matter and metric interact are nothing but the laws of gravitation; the g_{ik} in (2) are the components of the gravitational potential.–Whereas the the gravitational potentials are the components of an invariant quadratic differential form, *electromagnetic phenomena* are controlled by a four-potential, whose components ϕ_i are the components of an invariant *linear* differential form $\sum \phi_i dx_i$. However, both phenomena, gravitation and electricity, have remained completely isolated from one another up to now.

From recent publications of Levi-Civita [2], Hessenberg [3] and the author [4] it has become evident that a natural formulation of Riemannian geometry is based on the concept of infinitesimal parallel-transfer. If P and P' are two points connected by a curve then one can can transfer a vector from P to P' along the curve keeping it parallel to itself. However, the transfer of the vector from P to P' is, in general, not integrable i.e. the vector that is obtained at P' depends on the path. Integrability holds only for Euclidean ('gravitation-free') geometry. — But in the Riemannian geometry described above there is contained a residual element of rigid geometry—with no good reason, as far as I can see; it is due only to the accidental development of Riemannian geometry from Euclidean geometry. The metric (2) allows the magnitudes of two vectors to be compared, not only at the same point, but at any two arbitrarily seperated points. *A true infinitesimal geometry should, however, recognize only a principle for transferring the magnitude of a vector to an infinitesimally close point* and then, on transfer to an arbitrarily distant point, the integrability of the magnitude of a vector is no more to be expected than the integrability of its direction. On the removal of this inconsistency there appears a geometry that, surprisingly, when applied to the world, *explains not only the gravitational phenomena but also the electrical.* According to the resultant theory both spring from the same source, indeed *in general one cannot seperate gravitation and electromagnetism in an arbitrary manner.* In this theory *all physical quantities have a world-geometrical meaning; the action appears from the beginning as a pure number. it leads to an essentially unique universal law; it even allows us to understand in a certain sense why the world is four-dimensional.* —I shall first sketch the construction of the corrected Riemannian geometry without any reference to physics; the physical application will then suggest itself.

In a given coordinate system the coordinates dx_i of a point P' relative to an infinitesimally close point P are the components of the *infinitesimal translation* $\overrightarrow{PP'}$ — see (1). The change from one coordinate-system to another is expressed by the continuous transformation:

$$x_i = x_i(x_1^* x_2^* \dots x_n^*) \qquad (i = 1, 2, \dots, n)$$

which determines the connection between the coordinates of the same point in the different systems. For the components dx_i and dx_i^* of the same infinitesimal translation of the point P we then have the linear transformation

$$dx_i = \sum_k \alpha_{ik} dx_k^* \tag{3}$$

in which the the α_{ik} are the values of the derivatives $\partial x_i/\partial x_k^*$ at the point P. A (contravariant) *vector* at the point P has n numbers ξ^i as components in every coordinate system, and on transforming the coordinates, these numbers transform in the same way as the infinitesimal translations in (3). I shall call the set of all vectors at P the *vector-space* at P. It is 1. *linear or affine* i.e. it is invariant with respect to the multiplication of a vector by a number and the addition of two vectors, and 2. *metrical*: by means of the symmetric bilinear form (2) an invariant scalar product

$$\chi.\eta = \eta.\chi = \sum g_{ik}\chi^i\eta^k$$

is defined for each pair of vectors χ, η. However, according to our point of view *this form is determined only up to an arbitrary positive proportionality-factor*. If the manifold is described by the coordinates x_i only the ratios of the components g_{ik} are determined by the metric at P. Physically also, only the ratios of the g_{ik} have a direct physical meaning. For a given point P the neighbouring points P' which can receive light-signals from P satisfy the equation

$$\sum_{ik} g_{ik}dx_idx_k = 0$$

For the purpose of analytical representation we have to 1. choose a coordinate system and 2. in each point P determine the arbitrary proportionality-factor of the g_{ik}. Correspondingly each formula must have a double-invariance: 1. it must be *invariant with respect to arbitrary smooth coordinate transformations* 2. it must remain unchanged *when the g_{ik} are replaced by λg_{ik}* where λ is an arbitrary smooth function of position. Our theory is characterized by the appearance of this second invariance property.

An affine or linear map of the vector space at the point P onto the vector space at the point P^* is defined as the map $A \to A^*$ such that $\alpha\chi \to \alpha\chi^*$ and $\chi + \eta \to \chi^* + \eta^*$, where α is an arbitrary number. In particular the map is said to be is said to be a *similarity* map if the inner-product $\chi^*.\eta^*$ is proportional to the inner-product $\chi.\eta$ for all pairs of vectors χ and η. (Only this concept of similar maps has an objective meaning in our context; the previous theory allowed one to introduce the sharper concept of *congruent* maps.) The *parallel-transfer of a vector* at P to a neighbouring point P' is defined by the following two axioms:

1. The parallel transfer of the vectors at P to vectors at P' defines a similarity map.

2. If P_1 and P_2 are two neighbouring points to P and if the infinitesimal vectors $\overrightarrow{PP_2}$ and $\overrightarrow{PP_1}$ become $\overrightarrow{P_1P_{12}}$ and $\overrightarrow{P_2P_{21}}$, on parallel-transfer to P_2 and P_1 respectively then P_{12} and P_{21} coincide (commutativity).

The part of the first axiom that says that the parallel-transfer is an affine transformation of the vector space from P to P' is expressed analytically as follows: the vector ξ^i at $P = (x_1 x_2 \ldots x_n)$ is transferred to the vector

$$\xi^i + d\xi^i \quad \text{at} \quad P' = (x_1 + dx_1, x_2 + dx_2, \ldots, x_n + dx_n)$$

whose components are linear in ξ^i:

$$d\xi^i = -\sum_r d\gamma_r^i \xi^r \tag{4}$$

The second axiom requires that the $d\gamma_r^i$ are linear differential forms:

$$d\gamma_r^i = \sum_s \Gamma_{rs}^i dx_s,$$

whose coefficients have the symmetry property

$$\Gamma_{sr}^i = \Gamma_{rs}^i \tag{5}$$

If two vectors ξ^i, η^i at P are parallel-transferred to the vectors $\xi^i + d\xi^i$, $\eta^i + d\eta^i$ at P' the part of axiom 1 that goes beyond affinity to include similarity requires that

$$\sum_{ik} (g_{ik} + dg_{ik})(\xi^i + d\xi^i)(\eta^k + d\eta^k) \quad \text{and} \quad \sum_{ik} g_{ik}\xi^i \eta^k$$

are proportional. If we call the proportionality factor, which is infinitesimally close to unity, $(1 + d\phi)$ and define the lowering of indices in the usual manner as

$$a_i = \sum_k g_{ik} a^k$$

we then have

$$dg_{ik} - (d\gamma_{ki} + d\gamma_{ik}) = g_{ik} d\phi \tag{6}$$

From this it follows that $d\phi$ is a linear differential form:

$$d\phi = \sum_i \phi_i dx_i \tag{7}$$

If it is known, then the quantities Γ are determined by equation (6) or

$$\Gamma_{i,kr} + \Gamma_{k,ir} = \frac{\partial g_{ik}}{\partial x_r} - g_{ik}\phi_r$$

and the symmetry property (5). *The metrical connection of the space depends not only on the quadratic form (2)* (which is determined only up to a proportionality factor) *but on the linear form (7).* If, without changing coordinates, we replace g_{ik} by λg_{ik} the quantities $d\gamma_k^i$ remain unchanged, the $d\gamma_{ik}$ acquire a factor λ and dg_{ik} becomes $\lambda dg_{ik} + g_{ik}d\lambda$. Equation (6) then shows that $d\phi$ becomes

$$d\phi + \frac{d\lambda}{\lambda} = d\phi + d(\ln\lambda).$$

For the linear form $\phi_i dx_i$ the arbitrariness takes the form of an *additive total differential* rather than a proportionality factor that would be determined by a choice of scale. For the analytic representation of the geometry the forms

$$g_{ik}dx_i dx_k \qquad \phi_i dx_i \tag{8}$$

are on the same footing as

$$\lambda g_{ik}dx_i dx_k \quad \text{and} \quad \phi_i dx_i + d(\ln\lambda), \tag{9}$$

where λ is an arbitrary function of position. *The invariant quantity is therefore the anti-symmetric tensor with components*

$$F_{ik} = \frac{\partial\phi_i}{\partial x_k} - \frac{\partial\phi_k}{\partial x_i} \tag{10}$$

i.e. the form

$$F_{ik}dx_i\delta x_k = \frac{1}{2}F_{ik}\Delta x_{ik},$$

which depends bilinearly on two arbitrary translations dx and δx at the point P or, more precisely, on the surface-element

$$\Delta x_{ik} = dx_i\delta x_k - dx_k\delta x_i$$

determined by these two translations. The special case for which the magnitude of a vector at an arbitrary initial point can be parallel-transferred throughout the space in a path-independent manner appears when the g_{ik} can be chosen in such a way that the ϕ_i vanish. The Γ^i_{rs} are then nothing but the Christoffel 3-index symbols. The neccessary and sufficient condition for this to be the case is the vanishing of the tensor F_{ik}.

Accordingly, it is very suggestive to interpret ϕ_i as the electromagnetic potential and the tensor F as the *electromagnetic field.* Indeed, the absence of an electromagnetic field is the condition for the validity of Einstein's gravitational theory. If one accepts this interpretation one sees that electromagnetic quantities are such that their characterization by numbers in a given coordinate system is independent of the scale. In this theory one must adopt a new approach to the question of of scales and dimensions. Previously one spoke of a tensor being of second rank when, *after making an arbitrary choice of scale*, it was represented in every coordinate system by a matrix a_{ik} whose entries were the coefficients of an invariant bilinear form of two arbitrary independent infinitesimal translations

$$a_{ik}dx_i\delta x_k \tag{11}$$

Here we talk of a tensor when, having fixed a coordinate system and *making a definite choice of the proportionality factor of the* g_{ik}, the components a_{ik} are uniquely determined and indeed are determined in such a way that the form (11) is invariant with respect to coordinate transformations, but a_{ik} changes to $\lambda^e a_{ik}$ when g_{ik} changes to λg_{ik}. We say then that the tensor has *weight* e or, if a 'scale' l is assigned to the line-element ds, that it has dimension l^{2e}. The absolute invariant tensors are only those of weight zero. The field-tensor with the components F_{ik} is of this kind. According to (10) it satisfies the first system of Maxwell equations

$$\frac{\partial F_{kl}}{\partial x_i} + \frac{\partial F_{li}}{\partial x_k} + \frac{\partial F_{ik}}{\partial x_l} = 0$$

Once the concept of parallel-transfer is defined the geometry and tensor calculus is easily deduced.

a) *Geodesics*. Given a point P and a vector at P, the geodesic originating at P in the direction of this vector is obtained by continuously parallel-transferring the vector in its own direction. The differential equation for the geodesic takes the form

$$\frac{d^2x_i}{d\tau^2} + \Gamma^i_{rs}\frac{dx_r}{d\tau}\frac{dx_s}{d\tau} = 0$$

for a suitable choice of the parameter τ. (It cannot, of course, be interpreted as the line of shortest length since the concept of length along a curve is not meaningful.)

b) *Tensor Calculus*. For example, to obtain a tensor-field of rank 2 from a covariant tensor-field of rank 1 and weight zero and components f_i by differentiation, we take any vector ξ^i at the point P with coordinates x_i, construct the invariant $f_i\xi^i$ and compute its infinitesimal variation on parallel-transfer to a neighbouring point P' with coordinates $x_i + dx_i$. We obtain

$$\frac{\partial f_i}{\partial x_k}\xi^i dx_k + f_r d\xi^r = \left(\frac{\partial f_i}{\partial x_k} - \Gamma^r_{ik} f_r\right)\xi^i dx_k.$$

The quantities in brackets on the right-hand side are the components of a tensor of rank 2 and weight zero which has been derived from the field f in a fully invariant manner.

c) *Curvature*. To construct the analogue of the Riemann tensor consider the infinitesimal parallelogram consisting of the points P, P_1, P_2 and $P_{12} = P_{21}$. Since the points P_{12} and P_{21} coincide, it makes sense to compute the difference between the vectors obtained at this point by taking any vector $\xi = \xi^i$ and parallel-transferring it to P_{12} via P_1 and P_2 respectively. For its components one obtains

$$\Delta\xi^i = R^i_j \xi^j, \tag{12}$$

where the R^i_j are independent of the vector ξ but depend linearly on the surface-element spanned by the two infinitesimal transfers $\overrightarrow{PP_1} = (dx_i)$ and $\overrightarrow{PP_2} = (\delta x_i)$:

$$R^i_j = R^i_{jkl} dx_k \delta x_l = \frac{1}{2} R^i_{jkl} \Delta x_{kl}.$$

The curvature components R^i_{jkl}, which depend only on the point P, have the following two symmetry properties: 1. they change sign on permutation of the last indices k and l; 2. if one cyclically permutes the indices j, k, l and adds, the sum is zero. If the index i is lowered we obtain in R_{ijkl} the components of a covariant tensor of 4th rank and weight 1. One sees by inspection that R splits in an invariant manner into two parts

$$R^i_{jkl} = P^i_{jkl} - \frac{1}{2}\delta^i_j F_{kl} \qquad \delta^i_j = 1 \quad (i = j) \quad \delta^i_j = 0 \quad (i \neq j), \tag{13}$$

where P_{ijkl} is anti-symmetric in the indices i and j as well as k and l. Whereas the equations $F_{ik} = 0$ characterize the absence of an electromagnetic field i.e. a space in which the transfer of magnitude is integrable, one sees from (13) that $P^i_{jkl} = 0$ are the invariant conditions for the absence of a gravitational field i.e. for the parallel transfer of directions to be integrable. Only in Euclidean space is there neither electromagnetism nor gravitation.

The simplest invariant of a linear map like (12) that assigns a vector $\delta\xi$ to every ξ is the trace

$$\frac{1}{n} R^i_i.$$

For this we obtain from (13) the form

$$-\frac{1}{2} F_{ik} dx_i \delta x_k,$$

which we have already encountered. The simplest invariant that can be constructed from a tensor of the form $-F_{ik}/2$ is the square of its magnitude

$$L = \frac{1}{4} F_{ik} F^{ik}.$$

Since the tensor F has weight zero, L is clearly an invariant of weight -2.

If g is the negative determinant of the g_{ik} and

$$d\omega = \sqrt{g}dx_o dx_1 dx_2 dx_3 = \sqrt{g}dx$$

is the infinitesimal volume element, then, as is well-known, the Maxwell theory is determined by the electromagnetic Action, which is equal to the integral $\int L d\omega$ over an arbitrary volume of this simplest invariant, in such a way that for arbitrary variations of the g_{ik} and ϕ_i which vanish on the boundary we have

$$\delta \int L d\omega = \int \left(s^i \delta\phi_i + T^{ik} \delta g_{ik} \right) d\omega$$

where

$$s^i = \frac{1}{\sqrt{g}} \frac{\partial \left(\sqrt{g} F^{ik} \right)}{\partial x_k}$$

are the left-hand side of the Maxwell equations (on the right-hand side of which is the electromagnetic current) and the T^{ik} are the components of the energy-momentum tensor of the electromagnetic field. Since L is an invariant of weight -2 and the volume element an invariant of weight $\frac{n}{2}$ the integral $\int L d\omega$ then has a meaning only when the dimension is $n = 4$. *Thus in our context the Maxwell equations are possible only in 4 dimensions.* But in four dimensions the electromagnetic action is a pure number. Its magnitude in CGS units can, of course, only be determined when a computation based on our theory is applied to a physical problem such as the electron.

Passing on from Geometry to Physics, we have to assume, following the example of Mie's theory [5], that the whole set of natural laws is based on a definite integral-invariant, the action

$$\int W d\omega = \int \mathcal{W} dx \qquad (\mathcal{W} = W\sqrt{g})$$

in such a way that *the actual world is selected from the class of all possible worlds by the fact that the Action is extremal in every region* with respect to the variations of the g_{ik} and ϕ_k which vanish on the boundary of that region. W, the action-density, must be an invariant of weight -2. *The action is in any case a pure number*; in this way our theory gives pride of place to that part of atomic theory that is the most fundamental according to modern ideas: the action. The simplest and most natural Ansatz that we can make for W is

$$W = R^i_{jkl} R_i^{jkl} = |R|^2 \tag{14}$$

According to (13) this can be written as

$$W = |P|^2 + 4L$$

(At most the factor 4, by which the second [electrical] term is added, could be open to doubt). But even without specifying the action there are some general conclusions that we can draw. We shall show that: *just as* according to the researches of Hilbert [6], Lorentz [7], Einstein [8], Klein [9] and the author [10] *the four conservation laws of matter* (of the energy-momentum tensor) *are connected with the the invariance of the Action with respect to coordinate transformations*, expressed through four independent functions, the electromagnetic conservation law is connected with the new scale-invariance, expressed through a fifth arbitrary function. The manner in which the latter resembles the energy-momentum principle seems to me to be the strongest general argument in favour of the present theory—insofar as it is permissible to talk of justification in the context of pure speculation.

We set for an arbitrary variation which vanishes on the boundary

$$\delta \int \mathcal{W} dx = \int \left(\mathcal{W}^{ik}\delta g_{ik} + \mathbf{w}^i \delta\phi_i\right) dx \qquad \left(\mathcal{W}^{ik} = \mathcal{W}^{ki}\right) \tag{15}$$

The field equations are then

$$\mathcal{W}^{ik} = 0 \qquad \mathbf{w}^i = 0 \tag{16}$$

We can regard the first and second as the gravitational and the electromagnetic field equations respectively. The quantities defined by

$$\mathcal{W}_k^i = \sqrt{g} W_k^i, \qquad \mathbf{w}^i = \sqrt{g} w^i$$

are the mixed (contravariant) components of a tensor of weight -2 and rank 2 (1) respectively. In the system of equations (16) there are five superfluous equations corresponding to the invariance properties. This is expressed by the following five identities that hold for the left-hand sides:

$$\frac{\partial \mathbf{w}^i}{\partial x_i} \equiv \mathcal{W}_i^i; \tag{17}$$

$$\frac{\partial \mathcal{W}_k^i}{\partial x_i} - \Gamma_{kr}^s \mathcal{W}_s^r \equiv \frac{1}{2} F_{ik} \mathbf{w}^i. \tag{18}$$

The first is a result of scale-invariance. For if, in the transition from (8) to (9) we take $\ln\lambda$ to be an infinitesimal function of position we obtain the variation

$$\delta g_{ik} = g_{ik}\delta\rho, \qquad \delta\phi_i = \frac{\partial(\delta\rho)}{\partial x_i}.$$

For this (15) must vanish. If we express the invariance of the action with respect to coordinate transformations by an infinitesimal variation of the manifold [9][10] we obtain the identities

$$\left(\frac{\partial \mathcal{W}_k^i}{\partial x_i} - \frac{1}{2}\frac{\partial g_{rs}}{\partial x_k}\mathcal{W}^{rs}\right) + \frac{1}{2}\left(\frac{\partial \mathbf{w}^i}{\partial x_i}\phi_k - F_{ik}\mathbf{w}^i\right) \equiv 0,$$

which convert to (18) when $\frac{\partial \mathbf{w}^i}{\partial x_i}$ is replaced by $g_{rs}\mathcal{W}^{rs}$ according to (17). From the gravitational equations alone we obtain that

$$\frac{\partial \mathbf{w}^i}{\partial x_i} = 0 \tag{19}$$

and from the electromagnetic equations alone that

$$\frac{\partial \mathcal{W}_k}{\partial x_i} - \Gamma_{kr}^s \mathcal{W}_s^r = 0 \tag{20}$$

In Maxwell's theory $\mathbf{w}^i$ has the form

$$\mathbf{w}^i = \frac{\partial\left(\sqrt{g}F^{ik}\right)}{\partial x_k} - \mathbf{s}^i \qquad \left(\mathbf{s}^i = \sqrt{g}s^i\right),$$

where s^i is the four-current. Since the first part here identically satisfies (19) this yields the electromagnetic conservation law

$$\frac{1}{\sqrt{g}}\frac{\partial\left(\sqrt{g}s^i\right)}{\partial x_i} = 0.$$

In the same way $\mathcal{W}_k^i$ in Einstein's gravitational theory consists of two terms, the first of which identically satisfies equation (20), and the second of which is equal to the mixed energy momentum-tensor T_k^i multiplied by $\sqrt{g}$. In this way equation (20) leads to the four energy-momentum conservation equations. It is a completely analogous situation for our theory when we make the Ansatz (14) for the action. The five conservation laws can be eliminated from the field equations since they are obtained in two ways and thereby show that five of the field equations are superfluous.

For example for the Ansatz (14) the Maxwell equations read

$$\frac{1}{\sqrt{g}}\frac{\partial\left(\sqrt{g}F^{ik}\right)}{\partial x_k} = s^i \qquad \text{and} \qquad s_i = \frac{1}{4}\left(R\phi_i + \frac{\partial R}{\partial x_i}\right). \tag{21}$$

R denotes the invariant of weight -1 that is constructed from R^i_{jkl} by contracting i, k and j, l. The computation gives

$$R = R^* - \frac{3}{\sqrt{g}} \frac{\partial(\sqrt{g}\phi^i)}{\partial x_i} + \frac{3}{2}(\phi_i \phi^i),$$

where R^* denotes the Riemannian invariant constructed from the g^{ik}. In the static case, where the spacial components of the electromagnetic potential vanish and all quantities are independent of the time x_o, we must have, according to (21)

$$R = R^* + \frac{3}{2}\phi_o \phi^o = \text{const.}$$

But in a space-time region in which $R \neq 0$ one can quite generally, by suitable choice of the arbitrary scale, choose $R = \text{const} = \pm 1$. In time-dependent situations one must, however, expect to encounter surfaces where $R = 0$, which obviously play a singular role. R should not be used as an action since it is not of weight -2 (In Einstein's theory R^* is of this kind). This has the consequence that our theory leads to Maxwell's equations but not to Einstein's; instead of the latter we have fourth-order differential equations. But in fact it is not very probable that the Einstein gravitational field equations are strictly correct, particularly, since the gravitational constant contained in them is quite out of place with respect to the other natural constants, so that the gravitational radius of the mass and charge of an electron, for example, is of a completely different order of magnitude (about 10^{20} resp. 10^{40} times smaller) than the radius of the electron itself [11].

It was my intention to develop only the basis of the theory here. There arises the task of deriving the physical consequences of the Ansatz (14) and comparing them with experiment, in particular to see if they imply the existence of the electron and of other unexplained atomic phenomena. The problem is extraordinarily complicated from the mathematical point of view because it is out of the question to consider linear approximations. Since the neglect of non-linear terms in the interior of the electron is certainly not permissible, the linear equations obtained by neglecting them can have essentially only the trivial solution. I intend to return to these questions elsewhere.

Postscript. A Remark by Mr. A. Einstein Concerning the Above Work

If light-rays were the only means by which metrical relationships in the neighbourhood of a space-time point could be determined, there would indeed be an indeterminate factor left in the line-element ds (as well as in the g_{ik}). This ambiguity is removed, however, when measurements obtained through (infinitesimally small) rigid bodies and clocks are taken into account. A timelike

ds can be measured directly by a standard clock whose world-line is contained in ds.

Such a definition of the line-element ds would become illusory only if the assumptions concerning 'standard lengths' and 'standard clocks' was not valid in principle; this would be the case if the length of a standard rod (resp. speed of a standard clock) depended on its history. If this were really so in Nature, chemical elements with spectral-lines of definite frequency could not exist and the relative frequency of two neighbouring atoms of the same kind would be different in general. As this is not the case it seems to me that one cannot accept the basic hypothesis of this theory, whose depth and boldness every reader must nevertheless admire.

Author's reply

I thank Mr. Einstein for giving me the opportunity of answering immediately the objection that he raised. I do not believe, in fact, that he is correct. According to special relativity a rigid rod has always the same rest-length if it is at rest in an inertial frame, and, under the same circumstances, a standard clock has the same period in standard units (Michelson experiment, Doppler-effect). There is, however, no question of the clock measuring $\int ds$ when it is in arbitrary turbulent motion (as little as in thermodynamics an arbitrary fast and non-uniformly heated gas passes through only equilibrium states); it is certainly not the case when the clock (or atom) experiences the effect of a strongly varying electromagnetic field. In general relativity the most that one can say is: a *clock at rest* in a *static* gravitational field measures the integral $\int ds$ *in the absence of an electromagnetic field.* How a clock behaves in arbitrary motion in the common presence of arbitrary gravitational and electromagnetic fields can only be determined by the computation of the dynamics based on the physical laws. Because of this problematic behaviour of rods and clocks I have relied in my book *Raum-Zeit-Materie* only on the observation of light-signals for the measurement of the g_{ik} (P. 182ff.); in this way not only the ratios of these quantities but (by choice of a definite scale) even their absolute values can be determined *so long as the* Einstein *theory is valid.* The same conclusion has been reached independently by Kretschman [12].

According to the theory developed here, with a suitable choice of coordinates and the undetermined proportionality-factor, the quadratic form ds^2 is roughly the same as in special relativity, except in the interior of the atom, and the linear form is $= 0$ in the same approximation. In the case of no electromagnetic field (the linear form strictly $= 0$) ds^2 is exactly determined by the demand expressed in brackets (up to a *constant* proportionality-factor, which is also arbitrary in Einstein's theory; the same is true even for a static electromagnetic field). The most plausible assumption that can be made about a clock at rest in a static field is that it measures the ds which is normalized in this way; this assumption [13] has to be justified by an explicit dynamical calculation in both Einstein's theory and mine. In any case an oscillating system of definite structure that remains in a definite static field will behave in a definite way (the influence of a possibly turbulent history will quickly dissipate); I do not believe that my theory is in contradiction with this experimental situation (which is confirmed by the existence of chemical elements for the atoms). It is to be observed that the mathematical ideal of vector-transfer, on which the construction of the geometry is based, has nothing to do with the real situation regarding the movement of a clock, which is determined by the equations of motion.

The geometry developed here is, it must be emphasized, the true infinitesimal geometry. It would be remarkable if in nature there was realized instead an illogical quasi-infinitesimal geometry, with an electromagnetic field attached to it. But of course I could be on a wild-goose chase with my whole concept; we are dealing here with pure speculation; comparison with experiment is an understood requirement. For this the consequences of the theory must be worked out; I am hoping for assistance in this difficult task.

[1] B. Riemann, *Uber die Hypothesen, welche der Geometrie zugrunde liegen*, Math. Werke 2.Aufl. (Leipzig 1892) Nr. 13, S.272.
[2] T. Levi-Civita, Rend. Circ. Mat. Palermo **42** (1917).
[3] G. Hessenberg, Vectorielle Begrunding der Differentialgeometrie, Math. Ann. **78** (1917).
[4] H. Weyl, *Raum-Zeit-Materie* (Berlin 1918) §14.
[5] G. Mie, Ann Phys. **37**, **38**, **39** (1912/13). H. Weyl, *R-Z-M* §25.
[6] D. Hilbert, Die Grundlagen der Physik, 1 Mitt. Gott. Nachr. 20 Nov. 1915.
[7] H. A. Lorentz, in Vier Abhandlungen in den Versl. Kgl. Akad. van Wetensch., Amsterdam 1915/16.
[8] A. Einstein, Berl. Ber. (1916) 1111–1116.
[9] F. Klein, Gott. Nachr. 25 Januar 1918.
[10] H. Weyl, Ann. Phys. **54** (1917) 121–125.
[11] H. Weyl, *Zur Gravitationstheorie*, Ann. Phys. (1917) 133.
[12] E. Kretschman, *Uber den Physikalischen Sinn der Relativitatspostulate*. Ann. Phys. **53** (1917) 575.
[13] Part of whose experimental verification is still missing (red-shift of the spectral lines in the neighbourhood of large masses).

Postscript June 1955

This work was the beginning of the attempt to construct a 'unified field theory' which was taken up later by many others—without conspicuous success as far as I can see; as is well-known, Einstein himself was working at it until his death.

I completed the development of my theory in two papers [references omitted], further in the 4th and above all in the 5th edition of my book *Raum-Zeit-Materie*.

In this development I gave preference to another principle—first for formal reasons, then strengthened by an investigation of W. Pauli (Verh. dtsch. phys. Ges. *21* 1919).

The strongest argument for my theory seems to be this, that gauge-invariance corresponds to the conservation of electric charge in the same way that coordinate-invariance corresponds to the conservation of energy and momentum. Later the quantum-theory introduced the Schrödinger-Dirac potential ψ of the electron-positron field; it carried with it an experimentally-based principle of gauge-invariance which guaranteed the conservation of charge, and connected the ψ with the electromagnetic potentials ϕ_i in the same way that my speculative theory had connected the gravitational potentials g_{ik} with the ϕ_i, and measured the ϕ_i in known atomic, rather than unknown cosmological, units. I have no doubt but that the correct context for the principle of gauge-invariance is here and not, as I believed in 1918, in the intertwining of electromagnetism and gravity. Compare in this context my Essay [reference omitted]: *Geometry and Physics*.

2

Aftermath of Einstein's Gravitational Theory

As mentioned in the Introduction, it was Einstein's theory of gravitation that opened the way to the full understanding of gauge invariance and its geometrical significance. Einstein's original purpose, of course, was to explain the equivalence of gravitational and inertial mass, but in doing so he revolutionized the theory of gravitation by showing that it could be attributed entirely to the geometry of space-time. This aspect of Einstein's theory is so well-known that it needs no elaboration here. What is of interest for us is that Einstein's gravitational theory, which was based on Riemannian geometry, also provided the inspiration for the non-Riemannian geometry which is at the heart of gauge theory.

The first, and perhaps the most important, development came just a year later, when the mathematician Levi-Civita, inspired by what he called "la grandiosa concezione di Einstein" and the growing need to simplify the formalism, introduced the concept of parallel transfer [1].

What Levi-Civita realized was that the covariance of the Riemannian derivative and the Riemann tensor, namely,

$$\left(\nabla_\mu\right)^\alpha_\beta = \delta^\alpha_\beta \partial_\mu + \left\{ {\alpha \atop \mu\beta} \right\} \quad \text{and} \quad R^\alpha_{\mu\nu\beta} = [\nabla_\mu, \nabla_\nu]^\alpha_\beta, \tag{34}$$

respectively, where the $\{\}$ is the Christoffel connection defined as

$$\left\{ {\alpha \atop \mu\beta} \right\} = \frac{1}{2} g^{\alpha\sigma} \left(\partial_\mu g_{\beta\sigma} + \partial_\beta g_{\sigma\mu} - \partial_\sigma g_{\mu\beta} \right), \tag{35}$$

was due only to the transformation properties

$$\left\{ {\lambda \atop \mu\nu} \right\} \quad \rightarrow \quad \frac{\partial y^\alpha}{\partial x^\mu} \frac{\partial y^\beta}{\partial x^\nu} \frac{\partial x^\lambda}{\partial y^\gamma} \left\{ {\gamma \atop \alpha\beta} \right\} + \frac{\partial^2 y^\alpha}{\partial x^\mu \partial x^\nu} \frac{\partial x^\lambda}{\partial y^\alpha}, \tag{36}$$

of the Christoffel connection with respect to coordinate transformations, and not to the fact that it was derived from a metric. From this it was a short step to introducing connections as independent entities. Thus a connection came to be defined as any array of functions $\Gamma^\lambda_{\mu\nu}(x)$ (at first symmetric in the lower two indices) which satisfied the transformation law (36), and its properties were developed by Levi-Civita himself and by others such as Cartan [2] and Weyl [3].

In the same way as in (34) each connection Γ defines a covariant derivative $\nabla_\mu(\Gamma)$ and each such $\nabla_\mu(\Gamma)$ defines a Riemann tensor,

$$(\nabla_\mu)^\alpha_\beta = \delta^\alpha_\beta \partial_\mu + \Gamma^\alpha_{\mu\beta} \quad \text{and} \quad R^\alpha_{\mu\nu\beta} = [\nabla_\mu, \nabla_\nu]^\alpha_\beta. \tag{37}$$

Although the Riemann tensor defined with the covariant derivative ∇_μ has, in general, nothing to do with a metric, it retains the most important property of the metrical one: it vanishes if, and only if, Γ is (locally) trivial, i.e., if there exists a (local) coordinate system in which Γ is zero. When there is no metric, there is, of course, no question of raising or lowering indices, so $R^\alpha_{\mu\nu\beta}$ has one contravariant and three covariant indices. As one might expect from this, the only symmetries that it inherits from the metrical Riemann tensor is that it is anti-symmetric with respect to the indices μ and ν and that the sum of the tensors obtained by a cyclic permutation of the lower indices is zero.

Behind this algebraic formalism lay the geometrical idea of parallel transfer, by which a vector $v(x)$ is transferred along a curve C using not the infinitesimal increments $dv = (\partial_\mu v)dx^\mu$, but infinitesimal increments of the form

$$\delta v = \big(\nabla_\mu v\big) dx^\mu, \tag{38}$$

where the dx^μ are tangent to the curve and ∇_μ is the *covariant* derivative defined in (37).

In contrast to transfer with the ordinary derivative, the transfer (38) is independent of the coordinate system because, unlike the ∂v, the ∇v transform as tensors under general coordinate transformations,

$$\begin{aligned}
\partial_\mu v^\alpha(x) \quad &\rightarrow \quad \frac{\partial y^\nu}{\partial x^\mu}\frac{\partial x^\alpha}{\partial y^\beta}\partial_\nu v^\beta(y) + \frac{\partial y^\nu}{\partial x^\mu}\frac{\partial^2 x^\alpha}{\partial y^\nu \partial y^\beta} v^\beta \\
(\nabla v)^\alpha_\mu(x) \quad &\rightarrow \quad \frac{\partial y^\nu}{\partial x^\mu}\frac{\partial x^\alpha}{\partial y^\beta}(\nabla v)^\beta_\nu(y).
\end{aligned} \tag{39}$$

This means that the increments δv have an intrinsic geometrical meaning. They can also be used to define the Riemann tensor as the measure of the variation in a vector with respect to parallel transfer around an infinitesimal parallelogram, of sides dx and δx say, according to

$$\Delta v^\alpha = R^\alpha_{\mu\nu\beta} dx^\mu \delta x^\nu v^\beta. \tag{40}$$

This definition of the Riemann tensor is very intuitive and emphasizes its role as a commutator of two non-collinear transformations.

Since there is no concept of length in the non-metrical case, it is no longer possible to define a geodesic as the shortest curve between two points. But it may be defined as the curve obtained by the parallel transfer of a vector in its own direction, and (for a suitable choice of the curve parameter τ) the equation of the geodesic then turns out to be

$$\frac{d^2x^\lambda}{d\tau^2} + \Gamma^\lambda_{\mu\nu}\frac{dx^\mu}{d\tau}\frac{dx^\nu}{d\tau} = 0, \tag{41}$$

just as in the metrical case. Indeed, the covariance of the expression on the left-hand side of (41) was used by Levi-Civita to obtain the transformation properties of the Γ's.

The introduction of the general connection Γ produces more degrees of freedom because in general a (symmetric) connection has four times as many independent components as the corresponding metric tensor. More generally, in n dimensions a general (symmetric) connection has n times as many independent components as the metric tensor. As we shall see later, the introduction of Γ also permitted more flexibility in interpretation.

The concepts of parallel transport and general connections were introduced to physicists through Weyl's 1918 article, which will be discussed in the next chapter, and through his book *Space-Time-Matter* [4], which was published in 1919. It is interesting to quote Einstein's very positive reaction on receiving a copy of the proofs:

> Dear Colleague, I am reading the proofs of your book with admiration. It is like a master-symphony. Each little word is related to the whole and the structure is grandiose. The splendid method of deriving the Riemann tensor from parallel-transfer! How natural everything is!

Following Weyl, we shall refer to this book as RZM (from the German title *Raum-Zeit-Materie*).

The significance of the Levi-Civita-Weyl-Cartan development can hardly be overestimated. From the point of view of mathematics, it liberated Riemannian geometry from the metric and thus opened the way to a much more general concept of differential geometry, with the emphasis on differentiable manifolds and on their topological properties. This led to a sustained mathematical development which culminated about 1950 in the theory of of fiber-bundles mentioned in the Introduction and involved many other branches of mathematics such as analysis and cohomology. The history of that development is an interesting story in itself but lies well outside the scope of this book.

From the point of view of physics, the Levi-Civita-Weyl-Cartan development paved the way for a geometrical understanding of electromagnetism and of the weak and strong nuclear interactions and for understanding their common structure. As we shall see, the common feature of all the fundamental interactions is the *connection*. Indeed, for all the interactions except gravity, for which it is a derivative of the metric, the connection is actually the fundamental field. This situation may change, of course, as the interactions become more unified, but the importance of the connection is unlikely to diminish.

Within the theory of general symmetric connections, the Christoffel connections are distinguished by the fact that they are metric-compatible. This means that if a symmetric connection and a metric are assigned to a space the metric will be covariantly constant with respect to the connection if, and only if, the connection is the Christoffel one. In mathematical language,

$$\nabla_\lambda(\Gamma)g_{\mu\nu} = 0 \qquad \Leftrightarrow \qquad \Gamma^\lambda_{\mu\nu} = \left\{ {\lambda \atop \mu\nu} \right\}. \tag{42}$$

The equivalence of the two statements in (42) follows by using linear algebra. For example, the second equation follows from the first by cyclically permuting the indices in the first and subtracting the two resultant equations from it.

WEYL'S SYMMETRIC GENERALIZATION

An example of a more general connection than the Christoffel one is the connection used by Weyl in the 1918 paper reproduced here. This paper had a dual purpose. The first purpose was to construct what Weyl called a "true" infinitesimal geometry and the second was to use this geometry to construct a unified theory of gravitation and electromagnetism. We shall consider the geometry in this chapter and the physics in Chapter 3. Weyl's point was that standard Riemannian geometry is slightly defective because, although it purports to be infinitesimal, it contains a residue of rigid Euclidean geometry in the fact that the *magnitudes* of vectors, in contrast to their directions, are path-independent with respect to parallel transfer. In Weyl's opinion, a true infinitesimal geometry would not permit this anomaly, and he therefore proposed replacing the Christoffel connection by a (symmetric) connection of the form

$$\Gamma^\lambda_{\mu\nu} = \left\{ {\lambda \atop \mu\nu} \right\} + \frac{1}{2} g^{\lambda\sigma} \left(g_{\mu\sigma} v_\nu + g_{\sigma\nu} v_\mu - g_{\mu\nu} v_\sigma \right), \tag{43}$$

where $v_\mu(x)$ is a vector field. As shown by (42), such a connection is not compatible with the Riemannian metric, and to see what happens, one may take the Riemannian metric at a fixed point x_o and parallel-transfer it to all (connected) points of the space. One obtains in this way the non-local, non-Riemannian, metric $\hat{g}_{\mu\nu}$ where

$$\hat{g}_{\mu\nu} = e^{\int_{x_o}^{x} v_\lambda(y) dy^\lambda} g_{\mu\nu}(x). \tag{44}$$

Since the metric $\hat{g}$ is used to form inner products, it is easy to see that parallel transfer of a vector from x_1 to x_2 using the connection (43) changes the magnitude by a scale factor of the form

$$e^{\int_{x_1}^{x_2} v_\mu(x) dx^\mu}. \tag{45}$$

The important point is that in general the scale factor is *non-integrable*. That is, it depends on the path taken from x_1 to x_2, and this puts lengths on the same footing as directions. Indeed, the only case in which the scale factor is integrable is when v_μ is a gradient. To see this, one notes that in the path-independent case the integral around a closed loop is zero, and hence by the Stokes theorem,

$$e^{\int f_{\mu\nu} dx^\mu dx^\nu} = e^{\oint A_\mu dx^\mu} = 1 \quad \text{where} \quad f_{\mu\nu} = \partial_\mu v_\nu - \partial_\nu v_\mu, \tag{46}$$

where the loop is assumed to be local and therefore not affected by the global topology and the surface-integral is over any 2-dimensional surface that spans the loop. Since the loop is arbitrary, this implies that

$$f_{\mu\nu} = 0 \quad \text{or} \quad v_\mu = \partial_\mu \sigma(x) \qquad \text{locally}, \tag{47}$$

where $\sigma(x)$ is a scalar, as required.

It was in this paper of Weyl that the word *gauge* was introduced into differential geometry. It was quite appropriate since the scale factor attached to the metric changed the measurement of length and the word *gauge* was in common use for measurements of length, e.g., the width of railway tracks.

CARTAN'S ASYMMETRIC GENERALIZATION: TORSION

Another interesting generalization of the Christoffel connection, due to Cartan [2], was to remove the restriction that the connection be symmetric in the lower indices. Since the inhomogeneous term in the transformation law (36) is symmetric in these indices, one sees that the anti-symmetric part of the connection is actually a tensor $t^\lambda_{[\mu\nu]}$ of third rank and thus the Cartan connection $\tilde{\Gamma}^\lambda_{\mu\nu}$ is the sum of a symmetric connection and a third-rank tensor which is anti-symmetric in the lower indices

$$\tilde{\Gamma}^\lambda_{\mu\nu} = \Gamma^\lambda_{\{\mu\nu\}} + t^\lambda_{[\mu\nu]}, \tag{48}$$

where curly and square brackets denote symmetrization and anti-symmetrization, respectively. The anti-symmetric tensor t in (48) is called the torsion tensor. Thus where Weyl generalized the Christoffel connection by adding a symmetric term depending on a vector v, Cartan generalized it by adding an anti-symmetric term depending on the third-rank tensor t.

The Riemann tensor is still defined the same way as in (40) and the condition that the connection admits a covariantly constant metric is formally the same as before, namely,

$$\nabla_\lambda \left(\tilde{\Gamma}\right) g_{\mu\nu} = 0. \tag{49}$$

However, in contrast to the symmetric case, (49) cannot be solved for the connection in terms of the metric, but only in terms of the metric and the torsion. For this, one follows the previous procedure of permuting the indices and subtracting the two resultant equations from the original. One obtains in this way

$$\tilde{\Gamma}^\lambda_{\mu\nu} = \left\{ {\lambda \atop \mu\nu} \right\} + g^{\lambda\sigma}\theta_{\mu[\sigma\nu]}, \tag{50}$$

where the curly brackets are the usual Christoffel symbols (35) and

$$\theta_{\mu[\sigma\nu]} = t_{\mu[\sigma\nu]} + t_{\sigma[\mu\nu]} - t_{\nu[\mu\sigma]}, \tag{51}$$

is called the *contorsion* tensor. In the notation for θ, we have anticipated the fact that it is anti-symmetric in the last two indices. We also note for future reference that (51) can be inverted to recover the torsion tensor t from the contorsion tensor θ. Thus

$$\theta_{\mu[\sigma\nu]} = -\theta_{\mu[\nu\sigma]} \qquad \text{and} \qquad t_{\sigma[\mu\nu]} = \frac{1}{2}\left(\theta_{\nu[\mu\sigma]} - \theta_{\mu[\nu\sigma]}\right). \tag{52}$$

Naturally, the generalizations of the Christoffel connection to include scale and torsion are only the simplest examples of the choices that could be made. The general situation from the physical point of view is discussed in a clear and interesting manner by Schrödinger in his *Space-Time Structure* [5].

3

Generalizations of Einstein's Theory

Once Einstein's gravitational theory was established, there were a number of attempts to generalize it, in particular, to generalize it to include electromagnetism, which at the time was the only other interaction that was well-understood and regarded as fundamental. These attempts were not particularly successful, but three of them played an important role in the development of gauge theory and will therefore be briefly described.

WEYL'S THEORY OF GRAVITATION AND ELECTROMAGNETISM

The first generalization of Einstein's theory we wish to consider is that of Weyl. This proposal was contained in the 1918 paper that was mentioned in Chapter 2 and is reproduced in this volume. It was of critical importance for the development of gauge theory because it was the bridge by which the idea of non-metrical geometry was introduced to physicists.

In considering unified theories of gravitation and electromagnetism, Weyl suggested that since Riemannian geometry describes gravitation, a more general affine geometry might describe both gravitation and electromagnetism. The question was: which affine geometry? The Weyl geometry, described in Chapter 2, was an obvious candidate since it contained exactly one vector field in addition to the gravitational field. Furthermore, as we have seen, the geometry generated by this vector field is trivial (the phase factor becomes path-independent) if, and only if, the vector is a gradient, in exact analogy to the fact that the electromagnetic field becomes trivial if, and only if, electromagnetic potential is a gradient. Encouraged by this formal resemblance, Weyl proposed that the geometrical vector v should be identified as a multiple of the electromagnetic potential,

$$v_\mu(x) = \frac{e}{\gamma} A_\mu(x), \tag{53}$$

where γ is a constant. Then, just as gravitation can be thought of as being due to the path-dependence of vectorial *directions* induced by the metric-field $g_{\mu\nu}(x)$, electromagnetism could be thought of as being due to the path-dependence of vectorial *magnitudes* induced by the vector-field $v_\mu(x)$.

A consideration that convinced Weyl that he was on the right track was the fact that electromagnetic current conservation $\partial^\mu j_\mu(x) = 0$ would then follow from scale invariance in the same way that energy-momentum conservation $\partial^\mu T_{\mu\nu} = 0$ follows from Poincare invariance. Both conservation laws are, of course, special cases of Noether's theorem, but what impressed Weyl was that the conservation of momentum and charge, which are so different from the phenomenological point of view, would have a common geometrical basis.

Weyl's reasoning was obviously very original and deep and, as we shall see later, the general idea was correct. But the direct application to gravitational theory turned out to be unacceptable. This was pointed out in a note written by Einstein and published as a postscript at the end of the paper. The problem is that the lengths of measuring rods and the time measurements of clocks would be rescaled by the non-integrable factor $e^{\frac{\varepsilon}{\gamma}\int dx_\mu A^\mu}$ and would therefore depend on their history. This is in clear contradiction with the fact that atomic spectra (known very accurately at the time) depend only on the nature of the atoms and not on their histories. Weyl attempted to counter Einstein's objection in a further postscript, but, with hindsight, it is clear that Einstein was correct. Weyl's counter-argument was that, in the presence of the turbulent gravitational and electromagnetic fields that one would expect to find in the interior of atoms, the usual special-relativistic ideas concerning measuring rods and clocks break down, and to find out what actually happens one would have to solve the full non-linear matter-field equations. Thus the counter-argument was essentially a retreat from the usual ideas of measurement without any concrete proposals for their replacement. London, in the paper reproduced here, calls the counter-argument "a rather unclear re-interpretation of the concept of real scale, which robbed the theory of its immediate physical meaning and attraction." Many years later, Weyl ruefully admitted that he "was all too prone to mix up his mathematics with physical and philosophical speculation." London was nonetheless very impressed by the strength of Weyl's considerations and by his belief in his theory. He remarked that "it must have been an unusually strong metaphysical conviction that, in the face of such elementary physical evidence, prevented Weyl from abandoning the idea that Nature would have to make use of the beautiful mathematical possibility that was offered," and, as we shall see later, he played a major role in showing that Nature did, in fact, make use of this beautiful possibility. It simply used it in a different context. Thus London was among the first to realize that Weyl's general idea was correct and that he had erred only in its application.

In spite of the physical unacceptability of Weyl's original application, the 1918 paper represented a decisive step in the direction of modern gauge theory for two reasons. First, it showed for the first time how a geometrical significance could be ascribed to the electromagnetic field. Second, it was the bridge by which the concept of non-Riemannian connections was introduced to physicists. Weyl's ideas were later incorporated in the 1919 edition of his RZM book, where they also had an impact. It was through Weyl's 1918 paper that the word *gauge* was introduced into physics as well as geometry, and in both cases the word was quite appropriate since the rescaling of the metric tensor due to the path factor (45) could be interpreted as a change of length measurement and hence as a change of gauge. Unfortunately, when the scale factor was later changed to a phase factor, the word *gauge* was retained although it was no longer appropriate.

CARTAN'S THEORY OF GRAVITATION

The second generalization of Einstein's theory that is of some interest here is due to Cartan. In Cartan's theory the aim was to incorporate not electromagnetism but (integer) spin by means of torsion. Thus the manifold was still regarded as metrical and four-dimensional but the connection was taken to be the Christoffel connection plus the torsion tensor.

Postulating the existence of torsion is not sufficient to define a physical theory because torsion is an independent tensor field and thus in principle could interact with matter and gravitation in any arbitrary way and with any strength. To obtain a unique form of interaction Cartan used the minimal principle, i.e., he assumed that the interaction should be due only to the covariant derivative, just as in gravitation and electromagnetism. He further postulated that the Lagrangian should be the analogue of the Einsteinian one. Thus he proposed the action

$$\mathcal{L} = \int \sqrt{g}\left(R + kL(\bar{\psi} D\psi)\right), \tag{54}$$

where L reduces to the standard matter Lagrangian when $D \to \partial$, k is the gravitational constant, D is the covariant derivative formed with the full connection $\tilde{\Gamma}$, including torsion, and $\sqrt{g}R$ is the Riemann scalar density defined as

$$\begin{aligned} \sqrt{g}R &= \sqrt{g}g^{\mu\nu}R^{\sigma}_{\sigma\mu\nu} \\ &= \partial_\lambda\left(\sqrt{g}g^{\mu[\nu}\tilde{\Gamma}^{\lambda]}_{\mu\nu}\right) + \sqrt{g}g^{\mu\nu}\left(\tilde{\Gamma}^{\sigma}_{\lambda[\mu}\tilde{\Gamma}^{\lambda}_{\sigma]\nu} - \tilde{\Gamma}^{\sigma}_{\sigma\mu}t^{\lambda}_{\nu\lambda}\right), \end{aligned} \tag{55}$$

where t is the torsion tensor and $[\mu\sigma]$ denotes anti-symmetrization. Since the divergence vanishes on integration, the Lagrangian is quadratic in the connection, just as in the torsionless case. But it is also linear in the torsion. The field equations obtained by varying this Lagrangian with respect to the metric and torsion are

$$G_{\mu\nu} = kT_{\mu\nu} - D_\sigma U^{\sigma}_{\mu\nu} \quad \text{and} \quad T^{\lambda}_{\mu\nu} = kU^{\lambda}_{[\mu\nu]}, \tag{56}$$

where $T_{\mu\nu}$ and $U^{\lambda}_{[\mu\nu]}$ are the energy momentum and spin angular momentum densities, defined in the standard way as

$$T_{\mu\nu}(x) = \frac{\delta \mathcal{L}}{\delta g^{\mu\nu}(x)} \quad \text{and} \quad U^{\lambda}_{[\mu\nu]}(x) = \frac{\delta \mathcal{L}}{\delta t_{\lambda}^{[\mu\nu]}(x)}. \tag{57}$$

Because of the form of the Lagrangian, the field equations for the torsion derived from the action (54) are algebraic instead of differential. This has two consequences:

(a) The torsion, unlike the metric, is completely localized in the matter.

(b) The torsion can be eliminated algebraically by means of the field equations and the result substituted into the field equation for the metric to yield a modified metrical field equation of the form

$$G_{\mu\nu}(g,t) = kT_{\mu\nu} + k^2 S_{\mu\nu}, \tag{58}$$

where $S_{\mu\nu}$ is an extra spin contribution to the canonical energy-momentum tensor, and is quadratic in the spin angular momentum density U.

Unfortunately, Cartan's corrections to Einstein's theory are not testable under normal conditions. Because of the k^2 term in (58) the torsional corrections are of a higher order in k than the familiar Einsteinian corrections, such as the advance in the perihelion of mercury. A crude estimate may be obtained by noting that if $T_{\mu\nu}$ were of the order of the mass density ρ, then U would be of the order $h(\rho/m)$, where h is the unit of angular momentum (Planck's constant) and m a typical particle mass, and thus $S_{\mu\nu}$ would be of the order of $h^2\rho^2/m^2$. Thus the k^2 term in (58) would become significant only for $\rho \approx m^2/kh^2$, which is an extremely large density, of the order of 10^{50}grcm^{-3}. What is perhaps interesting is that it corresponds to a length $l = (m/\rho)^{\frac{1}{3}}$ which is close to the Planck length. An English translation of Cartan's original article is now available [1] and a full and clear description of the theory from a modern point of view can be found in the reviews by Hehl et al. [2].

KALUZA-KLEIN THEORY

The third proposal to generalize Einstein's geometry was made within the framework of Riemannian geometry. The idea was that since the increase of dimensions from 3 to 4 had led to Einstein's gravitational theory, a further increase from 4 to 5, using the 5-dimensional version of Einstein's theory, might lead to both gravitation and electromagnetism.

The first attempt at such a theory was made by Kaluza in the 1922 paper reproduced here and was purely classical. The meaning of the fifth coordinate, which Kaluza called x^o, was unclear, but he noted that this difficulty could be circumvented by taking the limit in which the x^o-dependence was zero, because this was compatible with the five-dimensional Einsteinian field equations (just as the static limit is compatible with 4-dimensional field equations) and did not trivialize the theory because the fifth dimension implied the existence of extra fields. Thus, if $g_{ab}(x)$ for $a, b = 0, 1, \ldots, 4$ denotes the 5-dimensional metric-field, with $\mu, \nu = 1 \ldots 4$ denoting the usual 4-dimensional space-time coordinates, there exist, in addition to the usual metric fields $g_{\mu\nu}(x)$, the five fields $g_{\mu o}(x)$, and $g_{oo}(x)$, and these do not vanish in the x^o-independent limit. Since $g_{\mu o}(x)$ is a four-vector, Kaluza identified it as a numerical multiple of the electromagnetic 4-potential,

$$g_{o\mu} = 2\alpha A_\mu \qquad \alpha = \text{constant}. \tag{59}$$

The Lagrangian chosen by Kaluza was the five-dimensional Einsteinian Lagrangian except that the scalar curvature was the 5-dimensional one. Thus

$$\begin{aligned} \mathcal{L} &= \int \sqrt{g} R \\ &= \int \partial_d \left(\sqrt{g} g^{ab} \left\{ {d \atop ab} \right\} \right) + \int \sqrt{g} g^{ab} \left\{ {c \atop d[a} \right\} \left\{ {d \atop c]b} \right\} \\ &= \int \sqrt{g} g^{ab} \left\{ {c \atop d[a} \right\} \left\{ {d \atop c]b} \right\}, \end{aligned} \tag{60}$$

where $[ac]$ denotes anti-symmetrization. Kaluza was gratified to find that, although the 5-dimensional Christoffel symbols contained all the derivatives of the gauge potential,

$$g_{oa} \left\{ {a \atop \mu\nu} \right\} = f_{\mu\nu} \equiv \partial_\mu A_\nu - \partial_\nu A_\mu \qquad g_{\nu\sigma} \left\{ {\sigma \atop \mu o} \right\} = h_{\mu\nu} \equiv \partial_\mu A_\nu + \partial_\nu A_\mu, \tag{61}$$

the Lagrangian of the theory contained only the $f_{\mu\nu}$.

He was further gratified to find that the 5-dimensional Einstein field equations split into the three equations

$$R_{\mu\nu} = 0 \qquad \nabla^\mu f_{\mu\nu} = 0 \qquad \nabla g_{oo} = 0, \tag{62}$$

where $f_{\mu\nu}$ is the usual electromagnetic field strength, ∇_μ is the 4-dimensional covariant derivative, and ∇ is the 4-dimensional Laplacian. Thus they reduce to the standard Einstein-Maxwell theory with the addition of a scalar field. Similarly the five dimensional geodesic equations $\delta u^a = 0$ split into the two equations

$$\frac{du^\lambda}{ds} = \Gamma^\lambda_{\rho\sigma} u^\rho u^\sigma + 2\alpha F^\lambda_\kappa u^o u^\kappa - g^{\lambda\rho}(\partial_\rho g_{oo}) u^o u^o \qquad \frac{du_o}{ds} = 2\alpha g_{oo} u^o, \tag{63}$$

which apart from the g_{oo} terms are just the gravitational and Lorentz force equations for the motion of a charged particle. Of course, in the usual manner that is used in general relativity, (63) could also be deduced from the conservation laws associated with the action (60) by considering world-tubes of matter in the infinitely thin limit. The reason that the 5-dimensional Einsteinian theory reduces so satisfactorily is that, as we shall see later, the reduction converts the 5-dimensional coordinate transformations into 4-dimensional coordinate transformations plus 1-dimensional gauge-transformations, and, for the quadratic Lagrangian, invariance with respect to these two sets of transformations forces it to be of the Einstein-Maxwell type.

Kaluza's theory was extended to cover the case of wave mechanics by Klein and Fock in 1926. Both papers are reproduced here. Klein's paper consists of two parts. In the first part he summarized the classical Kaluza theory and added some refinements. In the second part he modified it to bring it into line with Hamiltonian optics, Hamilton-Jacobi theory, de Broglie theory and wave mechanics. Fock's paper is rather similar and, as will be discussed later, has the additional merit of being one of the very first papers to stress the concept of gauge invariance in the framework of wave mechanics. In each paper the author followed the classic path used by Schrödinger [3] in the discovery of wave mechanics and threw some new light on the relationship between geometrical optics and wave-mechanics.

The greatest difficulty of the Kaluza-Klein-Fock theory is, of course, to explain *why* the fields should not depend on the fifth coordinate x^o, i.e., to find a natural dynamical mechanism for this independence. Some insight is gained by recasting the question in geometrical language because a weak x^o-dependence of the fields would follow from a relatively large curvature in the fifth direction. Thus x^o-independence could be interpreted as a "curling up" of the 5-dimensional space in one direction. But the problem of explaining why the 5th dimension should behave in this manner still remains.

The difficulty concerning the g_{oo} field raised by Kaluza, namely, how to identify it and how to interpret the geodesic for microscopic particles, for which the g_{oo} term is dominant rather than negligible, seems to have been less serious than he anticipated. Indeed, it stemmed mainly from his identification of the electromagnetic field with components of the metric tensor, and it disappears in the papers of Klein and Fock, who make the slightly different identification

$$A_\mu = g_{o\mu}/g_{oo}. \tag{64}$$

We shall see in the next section that this identification is the more natural one. With the identification (64) Klein and Fock obtained the *exact* Einstein-Lorentz equations of motion as the equations of null geodesics in the 5-dimensional space.

In spite of the fact that the standard Einstein-Maxwell-Lorentz theory and the wave-mechanical Maxwell-Lorentz theory could be derived from the Kaluza theory in the striking manner shown by the above three authors, the theory remained marginal because the unification it achieved was quite formal and made no new physical predictions. In particular, it left the gravitational and electromagnetic coupling constants unrelated.

DIMENSIONAL REDUCTION

The procedure by which higher-dimensional systems are reduced to lower-dimensional ones is called dimensional reduction. The reason that dimensional reduction is so powerful from the point of view of gauge theory is that it converts coordinate transformations in the full space into gauge transformations in the subspace. To see this, we first note that if the coordinates of the full space are denoted by $x_A = (x_\mu, x_a)$ where x_μ are the space-time coordinates, then the reduction restricts the full group of coordinate transformations to the space-time and "internal" subgroups

$$\text{(S-T)}\quad x^\mu \to \tilde{x}^\mu(x) \qquad \text{and} \qquad \text{(I)}\quad y^a \to \tilde{y}^a(x, y). \tag{65}$$

But since, for any vector field $v^A(x)$ on the space-time surface $y = 0$, the internal subgroup $y \to \tilde{y}$ in (65) leaves the space-time components $v^\mu(x)$ invariant and transforms the internal components $v^a(x)$ according to

$$v^a(x) \to R^a_b(x) v^b(x) \quad \text{where} \quad R^b_a(x) = \left(\frac{\partial \tilde{y}^a(x, y)}{\partial y^b}\right)_{y=0}, \tag{66}$$

we see that, from the space-time point of view, the internal subgroup may be viewed as a gauge group. Thus the internal transformations (I) in (65) have a dual interpretation, namely, as coordinate transformations in the full space or gauge transformations in the space-time subspace. We shall see later that their dual nature is reflected in the fact that that certain components of the Riemann tensor may be interpreted as components of curvature in the full space and as field strengths in space-time. The covariant derivative has a similar dual interpretation.

From the standard variation of the metric with respect to infinitesimal coordinate transformations $\delta x_A = \epsilon_A$, namely,

$$\delta g_{AB} = \nabla_A \epsilon_B + \nabla_B \epsilon_A = g_{AC} \partial_B \epsilon^C + g_{BC} \partial_A \epsilon^C + g_{AB,C} \epsilon^C, \tag{67}$$

we see that for the internal gauge transformations (I) in (65) we have in particular

$$\delta g_{ab} = g_{ab,c} \epsilon^c \qquad \delta g_{a\mu} = g_{ab} \partial_\mu \epsilon^b + g_{a\mu,b} \epsilon^b. \tag{68}$$

In the Kaluza case the infinitesimal internal transformations (I) in (65) reduce to

$$x^o \quad \rightarrow \quad \tilde{x}^o = x^o + \epsilon(x^\mu), \tag{69}$$

where $\epsilon(x)$ is an arbitrary function, and the variations of the metric (68) reduce to

$$\delta g_{oo} = 0 \qquad \delta g_{\mu o} = g_{oo}\partial_\mu \epsilon. \tag{70}$$

Thus the 4-vector $v_\mu = g_{\mu o}/g_{oo}$ transforms according to

$$v_\mu \quad \rightarrow \quad v_\mu + \partial_\mu \epsilon, \tag{71}$$

which suggests that it is this quantity which should be identified as a numerical multiple of the electromagnetic potential. This would agree with the identification of Klein and Fock and with that of Kaluza in the case where g_{oo} is assumed constant. Even though Kaluza did not assume that g_{oo} is constant, this is sufficient to explain the success of his scheme, since the final Lagrangian must be invariant with respect to (71), and in the limit of constant g_{oo}, this coincides with the usual electromagnetic gauge transformation. In particular, the quantities $f_{\mu\nu}$ in (60) and (61) are gauge-invariant in this limit whereas the quantities $h_{\mu\nu}$ are not, and this explains why the latter drop out.

In recent years there has been a revival of dimensional reduction [4]. This has happened for two reasons. First, the realization that phase transitions are the rule rather than the exception in physics has suggested phase transitions as a mechanism by which dimensional reductions could be produced. Thus, just as the phase transition associated with the spontaneous breakdown of electroweak symmetry produces particle masses of different orders of magnitude, one can envision that a phase transition could produce curvatures of different orders of magnitude. Second, the advent of string theory [5] has shown that the occurrence of higher-dimensional spaces, notably 10- and 26-dimensional spaces, may be quite natural (even inevitable). As is well-known, string theory, far from being a 4-dimensional theory in search of higher dimensions, is a higher-dimensional theory in search of a dimensional reduction. In this context the proposal that there be more dimensions than four does not seem so strange.

From the point of view of the development of gauge theory dimensional reduction is important for three seperate reasons:

First, as we have seen in (65), for example, there is a close connection between coordinate transformations in higher-dimensional spaces and gauge transformations in space-time.

Second, dimensional reduction played an important role in Fock's discovery of wave-mechanical gauge invariance and in London's discovery (see Chapter 4) of the true interpretation of Weyl's theory.

Third, the generalization of dimensional reduction to more than one extra dimension provided a reliable method for constructing the field strengths for *non-abelian* gauge fields at a time when no systematic method was known. As we shall see, it was used for just this purpose by Klein and Pauli in their attempts to formulate a geometrical theory of the strong nuclear interactions (see Chapter 7).

On the Unification Problem of Physics

by Th. Kaluza in Konigsberg

Sitzungsber. Preuss. Akad. Wiss. Berlin (1921) 966

In General Relativity a full description of the physics is obtained only if, in addition to the metric tensor $g_{\mu\nu}$, of the four-dimensional manifold, the electromagnetic four-potential q_μ is taken into account.

Although the residual duality of gravitation and electromagnetism does not diminish [the] beauty of this theory it nevertheless demands its replacement by a totally unified picture.

A surprisingly bold advance toward the solution of this problem, which is one of the favourite ideas of the human mind, was undertaken a few years ago by H. Weyl [1], who by a fundamental modification of the geometrical basis, introduced in addition to $g_{\mu\nu}$ a sort of metrical fundamental vector, which he interpreted as the electromagnetic potential: the complete metric thereby became the source of all physical phenomena.

Here we shall attempt to achieve the same result by a different method.

Apart from the difficulties of interpretation associated with the above profound theory of H. Weyl, it is conceivable that one could have an even more perfect realization of unification: gravitation and electromagnetism having as their source a single universal tensor.—I should like to show that such a close union of the two fundamental forces seems to be possible in principle.

The curl-form of the electromagnetic field-strength $F_{\mu\nu}$ and even more the obvious formal correspondence of the gravitational and electromagnetic field equations [2] suggest that the $\frac{1}{2}F_{\mu\nu} = \frac{1}{2}(q_{\mu,\nu} - q_{\nu,\mu})$ could somehow be truncated three-index symbols $\left[{i\lambda \atop \kappa}\right] = \frac{1}{2}(g_{i\kappa,\lambda} + g_{\kappa\lambda,i} - g_{i\lambda,\kappa})$. Indices separated by a comma will denote differentiation with respect to the corresponding coordinate. Latin indices will always run from 0 and Greek indices only from 1 to 4. If one considers this as a possibility one is led almost inevitably to an initially unattractive conclusion: that since in four-dimensions the only Christoffel symbols are the gravitional ones, this view of the $F_{\mu\nu}$ could be maintained only by introducing the rather strange idea of a fifth space-time dimension.

It is true that our previous physical experience contains hardly any hint of the existence of an extra dimension but nevertheless there is nothing to prevent us thinking of our four-dimensional space-time as part of an R_5; provided that we permit only four-dimensional variations of physical quantities to play a role, by regarding all derivatives with respect to the fifth space-time coordinate as either of a smaller order of magnitude or zero ("cylinder condition"). The fear that the effects of the fifth dimension might be cancelled by this assumption is groundless because of the way in which the coordinates are intertwined in the three-index symbols.

We consider therefore an R_5 and carry over to it the Einstein Ansätze; letting x^0 denote the extra coordinate to be added to the usual x^1 to x^4. Letting g_{rs} and Γ_{ikl} denote the metric tensor and three-index symbols symbols $\left[{ik \atop l}\right]$ for R_5 repectively, we find that, thanks to the cylinder condition, the Christoffel symbols reduce to:

$$2\Gamma_{\kappa\lambda\mu} = g_{\kappa\lambda,\mu} - g_{\lambda\mu,\kappa} - g_{\mu\kappa,\lambda} \qquad \text{(as before)}$$
$$2\Gamma_{o\kappa\lambda} = g_{o\kappa,\lambda} - g_{o\lambda,\kappa}, \qquad 2\Gamma_{\kappa\lambda o} = -(g_{o\kappa,\lambda} + g_{o\lambda,\kappa}),$$
$$2\Gamma_{oo\kappa} = g_{oo,\kappa}, \qquad 2\Gamma_{o\kappa o} = -g_{oo,\kappa} \qquad 2\Gamma_{ooo} = 0. \tag{1}$$

At first sight the result is not very encouraging. It is true that the $\Gamma_{o\kappa\lambda}$ appear as curls, but the ten components $\Gamma_{\kappa\lambda o}$, which according to the scheme should have an electromagnetic interpretation, appear to block the way. We continue nevertheless and, in order to make the $\Gamma_{o\kappa\lambda}$ proportional to the $F_{\kappa\lambda}$ we set:

$$g_{o\kappa} = 2\alpha q_\kappa, \qquad g_{oo} = 2h \tag{2}$$

so that the metric tensor for R_5 becomes essentially the gravitational metric tensor bordered by the electromagnetic four-potential; the role of the corner-component h is as yet undetermined. If the sums $q_{\kappa,\lambda} + q_{\lambda,\kappa}$ are, for brevity, denoted by $\Sigma_{\kappa\lambda}$, corresponding to the $F_{\kappa\lambda}$, one has

$$\Gamma_{o\kappa\lambda} = \alpha F_{\kappa\lambda} \qquad \Gamma_{\kappa\lambda o} = -\alpha\Sigma_{\kappa\lambda} \qquad \Gamma_{oo\kappa} = -\Gamma_{o\kappa o} = h_{,\kappa}. \tag{3}$$

The thirty-five Christoffel symbols of R_5, five of which are zero, are exhausted by the four-dimensional derivatives of h, the electromagnetic field $F_{\kappa,\lambda}$ and its auxiliary field $\Sigma_{\kappa\lambda}$. Furthermore, from the encompassing five-dimensional identities

$$(\Gamma_{ikl} + \Gamma_{kli} + \Gamma_{lik})_{,m} = \Gamma_{mik,l} + \Gamma_{mkl,i} + \Gamma_{mli,k} \tag{4}$$

we obtain, thanks to the cylinder condition, the known identities

$$F_{\kappa\lambda,\mu} + F_{\lambda\mu,\kappa} + F_{\mu\kappa,\lambda} = 0 \quad \text{and} \quad h_{,\kappa,\lambda} = h_{,\lambda,\kappa}. \tag{4a}$$

We now restrict, as usual, the choice of coordinates by letting $g = |g_{rs}| = -1$ and make the assumption (Approximation I) that the g_{rs} are close to their Euclidean values $-\delta_{rs}$. With $\Gamma^l_{ik} = -\left[{ik \atop l}\right] = -\Gamma_{ikl}$ the relevant components of the four-index symbols are then

$$\{\kappa\lambda;\, \mu o\} = \alpha F^{\lambda}_{\kappa,\mu} \qquad \{\kappa o;\, o\lambda\} = -h_{,\kappa,\lambda},$$
$$\{\kappa\lambda, oo\} = \{\kappa o, oo\} = \{oo, oo\} = 0. \tag{5}$$

Fortunately the auxiliary field of equation (3) no longer appears: The electromagnetic part of the curvature is due only to the derivatives of the field strengths. If one constructs finally the contracted tensor $R_{ik} = \{ir, rk\}$ one obtains, with the above assumptions:

$$R_{\mu\nu} = \Gamma^{\rho}_{\mu\nu,\rho} \quad \text{(as before)},$$
$$R_{o\mu} = -\alpha\nabla^{\nu}F_{\mu\nu},$$
$$R_{oo} = -\Delta h, \tag{6}$$

in standard notation. Thus the fifteen components of the curvature tensor decompose into: 1. the left-hand side of the old gravitational field-equations 2. the electromagnetic field-equations 3. a Poisson equation for the still uninterpreted field h. This provides a first justification for our Ansatz and for the hope of interpreting the gravitational and electromagnetic fields as components of a universal field.

For the material energy-momentum tensor that determines the right-hand side of the field-equations in R_5 we have in Approximation I:

$$T_{ik} = T^{ik} = \mu_o u^i u^k$$
$$(\mu_o = \text{rest-mass-density}, \quad u^r = \frac{dx^r}{ds}, \quad ds^2 = g_{lm}dx^l dx^m), \tag{7}$$

Since (for all three types of field-equation) we have $R_{o\mu} = -\kappa T_{o\mu}$ we see that the according to (6) the Maxwell equations require the conditions

$$I^{\mu} = \rho_o v^{\mu} = \frac{\kappa}{\alpha}T_{o\mu} = \frac{\kappa}{\alpha}\mu_o u^o u^{\mu}$$
$$(\rho_o = \text{charge density}, \quad v^{\rho} = \frac{dx^{\rho}}{d\sigma}, \quad d\sigma^2 = g_{\lambda\mu}dx^{\lambda}dx^{\mu}); \tag{8}$$

for the current. The space-time energy-momentum tensor is thus essentially bordered by the current density.

We continue the investigation under the assumption: u^o, u^1, u^2, $u^3 \ll 1$, $u^4 \sim 1$ (Approximation II). This requires in addition to small velocity a very small specific charge $\frac{\rho_o}{\mu_o}$ of the matter; because, since we then have $d\sigma^2 \sim ds^2$ and $v^\rho \sim u^\rho$, it follows from (8), if one also assumes that $\alpha = \sqrt{\frac{\kappa}{2}} = 3.06.10^{-14}$ In anticipation of the equations of motion, that

$$\rho_o = \frac{\kappa}{\alpha}\mu_o u^o = 2\alpha\mu_o u^o \ll \mu_o \tag{8a}$$

This equation shows first of all that in this case the electric charge may be interpreted essentially as a fifth component of momentum of the masses which move obliquely to the planes $x^o =$ constant. Thereby a further unification of two formerly unrelated concepts would appear to have been achieved.

Since, finally, in the Approximation II we have $T_{oo}, T_{11}, T_{22}, T_{33} \sim 0$ it follows from (7) that

$$T = g^{ik}T_{ik} = -T_{44} = -\mu_o, \tag{9}$$

and in particular for the usual form of the field equations of the first kind

$$R_{oo} = -R_{44} = \frac{\kappa}{2}\mu_o. \tag{10}$$

According to (6) the corner potential h manifests itself here essentially as a negative gravitational potential, while $\mathcal{G} = \frac{g_{44}}{2}$ has the usual meaning.

Now that the question of the important quantities in the field-equations has been settled in a satisfactory manner the question arises as to whether the 'geodesic' equations of motion in R_5, namely

$$\dot{u}^l = \frac{du^l}{ds} = \Gamma^l_{rs}u^r u^s \tag{11}$$

still represent the motion of charged matter in a gravitational and electromagnetic field in accordance with observation. In the approximation II this is no problem. Because of the interchangeability of ds and $d\sigma$ one obtains immediately from (3)

$$\bar{v}^\lambda = \frac{dv^\lambda}{d\sigma} = \Gamma^\lambda_{\rho\sigma}v^\rho v^\sigma + 2\alpha F^\lambda_\kappa u^o v^\kappa - h_{,\lambda}(u^o)^2 \tag{11a}$$

That is to say, because of the smallness of the term with $(u^0)^2$ for the ponderomotive force density, we have

$$\pi^\lambda = \mu_o\bar{v}^\lambda = \Gamma^\lambda_{\rho\sigma}T^{\prime\rho\sigma} + F^\lambda_\kappa I^\kappa \tag{12}$$

where we have taken $\alpha = \sqrt{\frac{\kappa}{2}}$. Thus the force automatically splits into gravitational and electrical parts of the usual form. Finally there remains for the o-component of (11) only:

$$\dot{u}^o = \alpha \Sigma_{44} = 2\alpha q_{4,4}, \tag{11b}$$

so that in the quasi-static situation that is required by Approximation II the quantity $\frac{d}{dx^4}\left(\frac{\rho_0}{\mu_0}\right) = 2\kappa q_{4,4}$ is of higher order (see (8a)). The required constancy of ρ_o would seem therefore to be guaranteed.

For the equations of motion also the auxiliary field remains irrelevant in this approximation.

Were Approximation II to correspond to reality the above would in its essential parts represent a satisfactory solution to the unification problem: a single potential-tensor (metric) would generate a universal field, which under ordinary conditions would split into a gravitational and electric field.

However, matter, in its fundamental constituents at least, is not weakly charged; in the words of H. Weyl its 'macroscopic placidity' stands in sharp contrast to its 'microscopic turbulence', and this is true in particular for the new coordinate x^o in the above scenario: for the electron or H-nucleus the quantity $\frac{\rho_o}{\mu_o}$ and with it the "velocity"-component is anything but small! In the form demanded by Approximation II the theory can describe at most macroscopic phenomena and the key question is whether it can be used for the above elementary particles.

If one tries to describe the motion of electrons by geodesics in R_5 one encounters immediately a difficulty [3] that threatens to destroy the whole structure. The problem is that, if one takes the earlier assumptions literally, the fact that $\frac{e}{m} = 1.77 \times 10^7$ (in lightseconds) means that the quantity u^o is so large that the last term in (11a), instead of disappearing, takes a value much greater than is observed experimentally and becomes the leading term. It is true that the transition to [a] large u^o requires modifications (the interchangeability of ds and $d\sigma$ is no longer valid) but nevertheless it would seem to be impossible to proceed in the old manner without some new hypotheses.

On the other hand I believe—with all due caution—that I see in the following direction a solution, which, if it succeeeded, would provide an even more satisfactory point of view. Since for reasonably small velocities of the matter and for arbitrary u^o the component R_{oo} remains approximately equal to $-R_{44}$, the two gravitational terms in (11a) can take the opposite sign, by suitable choice of the (up to now undetermined) reality character of x^o. It then seems that with the sacrifice of the, in any case rather questionable, gravitational constant κ there could be a reconciliation of the conflicting orders of magnitude, so that gravitation would appear as a sort of difference-effect. An attractive feature of this suggestion is the possibility it offers of attributing a statistical role to that constant. However, for the moment the consequences of this hypothesis cannot be foreseen; and of course there are other possibilities to consider. And threatening all universal hypotheses is the Sphinx of modern physics, the quantum theory.

In spite of all the physical and theoretical difficulties which are encountered in the above proposal it is hard to believe that the derived relationships, which could hardly be surpassed at the formal level, represent nothing more than a malicious coincidence. Should it sometime be established that the scheme is more than an empty formalism this would signify a new triumph for Einstein's General Theory of Relativity, whose suitable extension to five dimensions is our present concern.

[1] Sitzungsber. d. Berl. Akad. **1918** p. 465.
[2] See also H. Thirring. Phys. Ztschr. **19** p. 204.
[3] I wish to thank Mr. Einstein for drawing my attention to this discrepancy and for his interest in the above Ansätze.

[Note: The corner component g_{oo} of the metric tensor which is here denoted by h was denoted by Germanic g in Kaluza's paper. I have used h in order to avoid any confusion with the metric tensor and its determinant.—L. O'R.]

Quantum Theory and Five-Dimensional Relativity

by Oskar Klein in Copenhagen

Zeit. f. Physik 37 (1926) 895

In the following pages I should like to draw attention to a simple connection between the relation between the theory of gravitation and electromagnetism proposed by Kaluza [1] on the one hand and the approach to Quantum problems of de Broglie [2] and Schrödinger [3] on the other. The theory of Kaluza is based on the idea that the ten Einstein gravitational potentials g_{ik} and the four electromagnetic potentials ϕ_i can be connected through the cooefficients γ_{ik} of a line-element of a Riemannian space which contains a fifth dimension in addition to the usual four. The equations of motion of charged particles in electromagnetic fields then take the form of geodesic equations. If these equations are regarded as radiation equations , in which matter is regarded as a kind of wave-propagation, one arrives almost automatically at a second-order partial differential equation , which is a generalization of the usual wave-equation. If one restricts [oneself] to solutions of this equation for which the fifth dimension appears only harmonically, with a definite period associated with Planck's constant, one is led to the above-mentioned quantum-mechanical methods.

1. Five-Dimensional Relativity. I begin with a brief description of five-dimensional Relativity which is closely connected to Kaluza's theory but differs from it in certain points.

Let us consider a five-dimensional Riemannian line-element for which we assume there is a coordinate-independent meaning. We write it as

$$d\sigma = \sqrt{\sum \gamma_{ik} dx^i dx^k} \tag{1}$$

where the symbol Σ, here and throughout, denotes a summation over the double indices from 0 to 4 and $x^o \dots x^4$ denote the five coordinates of the space. The 15 quantities γ_{ik} are the covariant components of a five-dimensional symmetric tensor. To relate them to the usual quantities g_{ik} and ϕ_i of standard relativity theory we must of course make special assumptions. First , four of the coordinates, x^1, x^2, x^3, x^4 say, must always denote the standard four-space. Second, the quantities γ_{ik} should not depend on the fifth coordinate x^0. From this it follows that the permitted coordinate transformations are restricted to the following group [4]

$$x^o = x'^o + \psi_o(x'^1, x'^2, x'^3, x'^4)$$
$$x^i = \psi_i(x'^1, x'^2, x'^3, x'^4) \qquad (i = 1, 2, 3, 4) \tag{2}$$

Actually we should have written a constant times x'^o instead of x'^o in the first equation. The restriction of this constant to unity is, however, unimportant.

As is easily shown, γ_{00} is invariant with respect to the transformations (2). The assumption $\gamma_{oo} =$ constant is therefore permissible. This suggests that only the ratios of the γ_{ik} have a physical meaning, in which case this assumption is simply a possible convention. Since we have left the value of x^o undetermined we set

$$\gamma_{oo} = \alpha \tag{3}$$

One shows also that the following differentials

$$d\theta = dx^o + \frac{\gamma_{oi}}{\gamma_{oo}} dx^i, \tag{4}$$

$$ds^2 = \left(\gamma_{ik} - \frac{\gamma_{oi}\gamma_{ok}}{\gamma_{oo}} dx^i dx^k\right) \tag{5}$$

are invariant with respect to the transformations (2). In these expressions a summation over the doubled indices is understood. For such summations we wish, as usual, to drop the summation symbol. The differentials $d\theta$ and ds are connected to the line-element $d\sigma$ in the following manner:

$$d\sigma^2 = \alpha d\theta^2 + ds^2 \tag{6}$$

On account of the invariance of $d\theta$ and γ_{oo} it follows that for fixed x^o the four γ_{oi}, $i \neq o$, transform as the components of a four-vector. If x^0 is also transformed then a gradient of a scalar has to be added. This means that the quantities

$$\frac{\partial \gamma_{oi}}{\partial x^k} - \frac{\partial \gamma_{ok}}{\partial x^i}$$

transform like the covariant components F_{ik} of the electromagnetic field. Thus from the covariant point of view the quantities γ_{oi} behave like electromagnetic potentials. We therefore assume that

$$d\theta = dx^o + \beta\phi_i dx^i \tag{7}$$

that is

$$\gamma_{oi} = \alpha\beta\phi_i \qquad (i = 1, 2, 3, 4) \tag{8}$$

where β denotes a constant and the ϕ_i are defined so that in rectangular Galilean coordinates we have

$$(\phi_x, \phi_y, \phi_z) = A,$$
$$\phi_z = -cV \tag{9}$$

where A is the usual vector-potential, ϕ the usual scalar potential and c is the speed of light.

We should like to identify the differential form ds with the usual relativistic line-element. We therefore set

$$\gamma_{ik} = g_{ik} + \alpha\beta^2\phi_i\phi_k, \tag{10}$$

where the g_{ik} have been chosen so that in rectangular Galilean coordinates we have

$$ds^2 = dx^2 + dy^2 + dz^2 - c^2dt^2 \tag{11}$$

In this way the γ_{ik} are identified with known quantities. The problem now is to construct field equations for the γ_{ik} which, in sufficient approximation, will lead to the usual relativistic field equations for the g_{ik} and ϕ_i. We do not wish to consider the details of this difficult problem, but only show that the usual field equations are easily combined from the point of view of five-dimensional geometry. We construct the invariants

$$P = \sum \gamma^{ik}\left(\frac{\partial\left\{{i\mu \atop \mu}\right\}}{\partial x^k} - \frac{\partial\left\{{ik \atop \mu}\right\}}{\partial x^\mu} + \left\{{i\mu \atop \nu}\right\}\left\{{k\nu \atop \mu}\right\} - \left\{{ik \atop \mu}\right\}\left\{{\mu\nu \atop \nu}\right\}\right) \tag{12}$$

where γ^{ik} are the contravariant components of the five-dimensional metric tensor and the $\left\{{rs \atop i}\right\}$ are the Christoffel symbols:

$$\left\{{rs \atop i}\right\} = \frac{1}{2}\sum \gamma^{i\mu}\left(\frac{\partial\gamma_{\mu r}}{\partial x^s} + \frac{\partial\gamma_{\mu s}}{\partial x^r} - \frac{\partial\gamma_{rs}}{\partial x^\mu}\right) \tag{13}$$

In the expression for P we assume that all the quantities are independent of x^o and that $\gamma_{oo} = \alpha$. Let us now consider the integral

$$J = \int P\sqrt{-\gamma}dx^o dx^1 dx^2 dx^3 dx^4, \tag{14}$$

taken over a closed region of the five-dimensional space, where γ denotes the determinant of γ_{ik}. We construct δJ by variation of the quantities γ_{ik} and $\frac{\partial\gamma_{ik}}{\partial x^l}$

where the variations vanish on the boundary and α is regarded as a constant. The variational principle

$$\delta J = 0 \tag{15}$$

leads to the following equations

$$R^{ik} - \frac{1}{2} g^{ik} R + \frac{\alpha\beta^2}{2} S^{ik} = 0 \qquad (i, k = 1, 2, 3, 4) \tag{16a}$$

$$\frac{\partial \sqrt{-g} F^{i\mu}}{\partial x^\mu} = 0 \qquad (i, k = 1, 2, 3, 4) \tag{16b}$$

where R is the Einstein scalar curvature, R^{ik} the contravariant components of the Einstein curvature tensor, g^{ik} the contravariant components of the metric tensor, S^{ik} the contravariant components of the electromagnetic energy-tensor, g the determinant of g_{ik} and finally $F^{i\mu}$ the contravariant components of the electromagnetic field-tensor. If we set

$$\frac{\alpha\beta^2}{2} = \kappa \tag{17}$$

where κ is the usual Einstein gravitational constant, we see that equations ($16a$) and ($16b$) are in fact identical with the matter-free field equations for the gravitational field and the generalized Maxwell equations respectively [5].

If we restrict ourselves to the usual scheme for handling the interaction of the gravitational and electromagnetic fields with matter we can obtain the corresponding equations in the presence of matter in a similar way. We replace P in (14) by

$$P + \kappa \sum \gamma_{ik} \Theta^{ik}.$$

To define the Θ^{ik} we first consider the tensor corresponding to an electron or a hydrogen nucleus:

$$\theta^{ik} = \frac{dx^i}{dl} \frac{dx^k}{dl} \tag{18}$$

where dx^i denotes the infinitesimal change of position of the particle and dl denotes a certain invariant differential. The Θ^{ik} is taken to be the sum of the θ^{ik} in a unit volume. We arrive at equations which are identical to those of the usual type if we set

$$v_o \frac{d\tau}{dl} = \pm \frac{e}{\beta c}, \tag{19}$$

$$\frac{d\tau}{dl} = \begin{cases} \sqrt{M} \\ \sqrt{m} \end{cases} \tag{20}$$

where generally

$$v^i = \sum \gamma_{i\mu} \frac{dx^\mu}{dl} \tag{21}$$

are the covariant components of the five-dimensional velocity, where

$$v^i = \frac{dx^i}{dl} \tag{22}$$

Furthermore, e denotes the quantum of electric charge, M and m the masses of the hydrogen nucleus and electron respectively, with the upper upper and lower signs for the nucleus and electron respectively. Also

$$d\tau = \frac{1}{c}\sqrt{-ds^2}$$

is the differential of the proper time.

The field equations imply the equations of motion for material particles and the continuity equations in the usual way. The computations which lead to these can easily be summarized from our point of view. As is easily seen our field equations are equivalent to the following 14 equations:

$$P^{ik} - \frac{1}{2}\gamma^{ik}P + \kappa\Theta^{ik} = 0 \tag{23}$$

($i, k = 0, 1, 2, 3, 4$, but not both zero), where the P^{ik} are the contravariant components of the contracted five-dimensional curvature tensor (corresponding to the R^{ik}). The relevant equations then follow by taking the divergence of (23). From this it follows that the electrically charged particles move on five-dimensional geodesics satisfying [6] conditions (19) and (20) . As is easily seen, these conditions are compatible with the geodesic equations because x^o does not appear in the γ_{ik}.

It should be recalled that there are not really sufficient grounds for the exact validity of Einstein's field equations. Nevertheless it should be not without interest that all 14 field equations can be unified in such a simple way by the Kaluza theory.

2. The Wave-equation of Quantum Theory. We now procede to establish the connection between the theory of stationary states and the deviations from mechanics that appear in the new quantum mechanics and five-dimensional relativity theory. Let us consider for this purpose the following differential equation [which is] related to our five-dimensional space and can be regarded as a generalization of the wave-equation:

$$\sum a^{ik}\left(\frac{\partial^2 U}{\partial x^i \partial x^k} - \sum \left\{ {ik \atop r} \right\} \frac{\partial U}{\partial x^r}\right) = 0 \tag{24}$$

Here the a^{ik} denote the contravariant components of a five-dimensional symmetric tensor which are certain functions of the coordinates. Equation (24) is independent of the coordinates.

Let us consider first a wave propagation determined by (24) and corresponding to the limiting case of geometrical optics. We arrive at it by setting

$$U = Ae^{i\omega\Phi} \tag{25}$$

where we assume that ω is so large that we need only consider the terms proportional to ω^2. We then obtain

$$\sum a^{ik} \frac{\partial \Phi}{\partial x^i} \frac{\partial \Phi}{\partial x^k} = 0 \tag{26}$$

an equation which corresponds to the Hamilton-Jacobi partial differential equation of mechanics. If we set

$$p_i = \frac{\partial \Phi}{\partial x^i} \tag{27}$$

then, as is known, the differential equations for the rays may be written in the following Hamiltonian form

$$\frac{dp_i}{-\frac{\partial H}{\partial x^i}} = \frac{dx^i}{\frac{\partial H}{\partial p_i}} = d\lambda \tag{28}$$

where

$$H = \frac{1}{2} \sum a^{ik} p_i p_k \tag{29}$$

From (26) it follows that

$$H = 0 \tag{30}$$

An alternative, Lagrangian, representation of these equations is obtained from the fact that the rays can be regarded as the null geodesics of the differential form

$$\sum a_{ik} dx^i dx^k$$

where the a_{ik} denote the reciprocals of the a^{ik}, that is

$$\sum a_{i\mu} a^{k\mu} = \delta_i^k = \begin{cases} 1, & i = k \\ 0, & i \neq k \end{cases} \tag{31}$$

If we now set

$$\sum a_{ik} dx^i dx^k = \mu (d\theta)^2 + ds^2 \tag{32}$$

then we can arrange by a suitable choice of the constant μ that our ray equations are identical with the equations of motion of charged particles. To see this set

$$L = \frac{1}{2}\mu\left(\frac{d\theta}{d\lambda}\right)^2 + \frac{1}{2}\left(\frac{ds}{d\lambda}\right)^2, \tag{33}$$

from which we have

$$p_o = \frac{\partial L}{\partial \frac{dx^o}{d\lambda}} = \mu\frac{d\theta}{d\lambda} \tag{34}$$

and

$$p_i = \frac{\partial L}{\partial \frac{dx^i}{d\lambda}} = u_i\frac{d\tau}{d\lambda} + \beta p_o \phi_i \qquad (i = 1, 2, 3, 4) \tag{35}$$

where $u_1 \ldots u_4$ denote the covariant components of the usual velocity vector.

The ray equations now read

$$\frac{dp_o}{d\lambda} = 0 \tag{36a}$$

$$\frac{dp_i}{d\lambda} = \frac{1}{2}\frac{\partial g_{\mu\nu}}{\partial x^i}\frac{dx^\nu}{d\lambda}\frac{dx^\nu}{d\lambda} + \beta p_o\frac{\partial \phi_\mu}{\partial x^i}\frac{dx^\mu}{d\lambda} \qquad (i = 1, 2, 3, 4) \tag{36}$$

From

$$\mu d\theta^2 + ds^2 = \mu d\theta^2 - c^2 d\tau^2 = 0$$

we have

$$\mu\frac{d\theta}{d\tau} = c\sqrt{\mu} \tag{37}$$

Since from (34) and (36a) $\frac{d\theta}{d\lambda}$, and hence also $\frac{d\tau}{d\lambda}$, is constant, we can choose λ so that

$$\frac{d\tau}{d\lambda} = \begin{cases} M \\ m \end{cases} \tag{38}$$

for the H-nucleus and the electron respectively. In order to obtain the usual equations of motion we must also assume

$$\beta p_o = \pm\frac{e}{c} \tag{39}$$

for the H-nucleus and electron respectively. From (37) we then have

$$\mu = \frac{e^2}{\beta^2 M^2 c^4} \qquad \text{and} \qquad \mu = \frac{e^2}{\beta^2 m^2 c^4}, \tag{40}$$

for the respective cases. Equations (35) and (36) are then in complete agreement with the equations for the motion of charged particles in gravitational and electromagnetic fields. In particular the quantities p_i defined in (35) are identical with the usual generalized momenta, a circumstance that is important for the following considerations. Since we can chose β arbitrarily we choose

$$\beta = \frac{e}{c} \tag{41}$$

from which we easily obtain

$$p_0 = \pm 1 \tag{39a}$$

and

$$\mu = \frac{1}{M^2c^2} \qquad \text{and} \qquad \frac{1}{m^2c^2} \tag{40a}$$

for the H-nucleus and electron respectively.

One sees that in (37) we must take the positive and negative signs of the square root for the nucleus and electron respectively. This is unsatisfactory. The fact, however, that with a single value of μ one obtains two different classes of rays, which are related to each other in much the same way as positively and negatively charged particles, could be construed as a hint that it might be possible to change the wave-equation in such a way that that the equations of motion of both kinds of particle could be derived from a single value of the coefficient. We do not wish to pursue this question further but proceed to consider in more detail the wave-equation derived from (32) in the case of the electron.

Since for the electron it was assumed that $p_o = -1$ we should, according to (27) set

$$\Phi = -x^o + S(x^1, x^2, x^3, x^4). \tag{42}$$

We obtain de Boglie's theory if we seek standing waves which are compatible with the wave-equation and correspond to a definite value of ω, and assume that the wave propagation proceeds according to the laws of geometrical optics. For this we need the well-known conservation law for the phase, which follows immediately from (28) and (30). We have, namely

$$\frac{d\Phi}{d\lambda} = \sum \frac{\partial \Phi}{\partial x^i}\frac{\partial x^i}{\partial \lambda} = \sum p_i \frac{\partial H}{\partial p_i} = 2H = 0. \tag{43}$$

This means that the phase is carried along by the wave. Let us consider now the simple case in which Φ decomposes into two terms, one of which depends only on a single coordinate, x say, which oscillates periodically in time. It will then be possible to have a standing wave which is characterized by the fact that, after a period of x, a harmonic wave that is represented by (25) at a given moment coincides in phase with the wave which is obtained from the same solution (25) by inserting new values of x^o, x^1, x^2, x^3. On account of the conservation of phase, the condition for this is simply

$$\omega \oint p dx = n.2\pi, \tag{44}$$

where n is an integer. If we set

$$\omega = \frac{2\pi}{h}, \tag{45}$$

where h denotes Planck's constant, we obtain the usual quantum condition for a separable coordinate. A similar result holds, of course, for an arbitrary periodic system. The usual quantum theory of periodic systems completely corresponds therefore with the treatment of interference effects through the assumption that the waves propagate according to the laws of geometrical optics. It should also be emphasized that on account of (42) the relations (44), (45) are invariant with respect to the coordinate transformations (2).

Let us also consider equation (24) in the case that ω is not so large that we need only take the quadratic terms into account. We restrict ourselves to the simple case of the electrostatic field. Then in Cartesian coordinates we have

$$\begin{aligned} d\theta &= dx^o - eV dt \\ ds^2 &= dx^2 + dy^2 + dz^2 - c^2 dt^2. \end{aligned} \tag{46}$$

Thus

$$H = \frac{1}{2}(p_x^2 + p_y^2 + p_z^2) - \frac{1}{2c^2}(p_t + eVp_o)^2 + \frac{m^2c^2}{2}p_o^2. \tag{47}$$

In equation (24) we can now neglect the terms proportional to the $\left\{ {ik \atop r} \right\}$ since according to (17) the Christoffel symbols are in this case smaller by a factor of κ. We therefore have [7]

$$\Delta U - \frac{1}{c^2}\frac{\partial^2 U}{\partial t^2} - \frac{2eV}{c^2}\frac{\partial^2 U}{\partial t \partial x^o} + \left(m^2c^2 - \frac{e^2V^2}{c^2}\right)\frac{\partial^2 U}{\partial x^{o2}} = 0. \tag{48}$$

Since V depends only on x, y, z we can, consistently with (42) and (45), make the Ansatz

$$U = e^{-2\pi i\left(\frac{x^o}{h} - \nu t\right)}\psi(x, y, z). \tag{49}$$

for U. When this is substituted into (48) we obtain

$$\Delta\psi + \frac{4\pi^2}{c^2h^2}[(h\nu - eV)^2 - m^2c^4]\psi = 0. \tag{50}$$

If we set

$$h\nu = mc^2 + E, \tag{51}$$

we obtain the Schrödinger equation [8], whose stationary oscillations give values of E which agree with the energy values of the Heisenberg quantum theory. One sees that in the limit of geometrical optics E is equal to the usual mechanical energy. According to (51) the frequency condition means that, as mentioned by Schrödinger, the light-frequencies associated with the system are equal to the differences of the frequencies ν.

3. Concluding Remarks. As in the work of de Broglie, the above considerations are the result of an attempt to use the analogy between Mechanics and Optics which appears in Hamilton's theory, in order to obtain a deeper understanding of quantum phenomena. That this analogy has a real physical meaning is indicated by the similarity of the conditions for stationary states of atomic systems and the inteference phenomena of optics. Of course known concepts such as point-charges and point-matter are foreign to classical field theory. The hypothesis has indeed often been made that material particles are to be considered as special solutions of the equations that determine the gravitational and electromagnetic fields. It is suggestive to relate the above-mentioned analogy to this point of view. Since according to this hypothesis it is not so strange that the movement of material particles would show similarities with wave-propagation. The relevant analogy is nevertheless incomplete so long as one considers wave-propagation in only four dimensions. This is seen already in the variable velocity of material particles. But if one thinks of the observed motion as a kind of projection into space-time of a wave-propagation which takes place in five-dimensions, the analogy becomes complete, as we have seen. In mathematical terms this means that the Hamilton-Jacobi equation should be considered as the equation for the characteristics of a five-dimensional rather than a four-dimensional wave-equation. In this way one is led to the theory of Kaluza.

Although the introduction of a fifth dimension in our physical considerations may seem rather strange at first sight, a radical modification of the geometry underlying the field equations is suggested by quantum theory for a different reason. It is known to be less and less probable that the quantum phenomena will admit an intrinsic space-time description, whereas the possibility of representing these phenomena by a system of five-dimensional field equations cannot be rejected [9] a priori. Whether the indicated possibilities contain some truth only the future can tell. In any case it should be emphasized that the approach given in this work should be considered as only provisional with respect to both the field-equations and the theory of stationary states. This can be seen from the schematic treatment of matter discussed at the end of §1, as well as in the circumstance that two types of charged particles are treated by different equations of the Schrödinger type. There is also left open the question as to whether the fourteen potentials are sufficient to describe the physical phenomena or whether the Schrödinger method requires the introduction of a new state-variable.

[1] Th. Kaluza, Sitzungsber. d. Berl. Akad. 1921, S.9662)
[2] L. de Broglie, Ann. d. Phys. (10) **3**, 22, 1925. Theses, paris 1924)
[3] E. Schrödinger, Ann. Phys. **79**, 361 and 489, 1926)
[4] See H. A. Kramers, Proc. Amsterdam **23**, No. 7, 1922, where a simple proof of the invariance of $d\theta$ and ds^2, reminiscent of the following considerations, is given.
[5] See H. Pauli, Relativitatstheorie, pages 719 and 724
[6] The special values of $\frac{d\tau}{dl}$ are, of course, of no significance in this context. What is important is only that $\frac{d\tau}{dl}$ is constant.
[7] Apart from the appearance of x^o, which plays no role in the application, this equation differs from the Schrödinger equation in the way in which in (48) the time appears. As support for this kind of quantum equation one might mention that in the case that V depends harmonically on the time, it possesses solutions which have the same relation to [the] Kramers dispersion relations as the Schrödinger solutions to the quantum theory of spectral lines. This is easily seen by a perturbative computation. I am grateful to Dr. W. Heisenberg for this remark.
[8] E. Schrödinger, *loc. cit.*
[9] Remarks of this kind, made by Prof. Bohr on several occasions, have had a decisive influence on my producing this note.

The considerations presented in this paper were the result of work done at the Physical Institute of [the] University of Michigan at Ann Arbor as well as this Institute. I should like to express my warmest thanks to Professors H. M. Randall and N. Bohr.

On the Invariant Form of the Wave and Motion Equations for a Charged Point-Mass [1]

by V. Fock in Leningrad

Zeit. f. Physik 39 (1927) 226

The Schrödinger wave-equation and the equations of motion are written as an invariant Laplace equation and the equation of a geodesic in five-dimensional space respectively. The superfluous fifth coordinate is closely related to the linear differential form of the electromagnetic potential.

In an article still to be published H. Mandel [2] concerns himself with the concept of five-dimensional space in order to view the gravitational and electromagnetic fields from a unified point of view. The introduction of the fifth coordinate seems to us very suitable for the invariant formulation of both the Schrödinger wave-equation and the mechanical equations of motion.

1. Special Relativity Theory

The Lagrange function for the motion of a charged point-mass is, in readily understood notation,

$$L = -mc^2\sqrt{1-\frac{v^2}{c^2}} + \frac{e}{c}\mathcal{A}.v - e\phi = 0, \tag{1}$$

and the corresponding Hamilton-Jacobi equation takes the form

$$(\operatorname{grad} W)^2 - \frac{1}{c^2}\left(\frac{\partial W}{\partial t}\right)^2 - \frac{2e}{c}\left(\mathcal{A}.\operatorname{grad} W + \frac{\phi}{c}\frac{\partial W}{\partial t}\right) \\ + m^2c^2 + \frac{e^2}{c^2}(\mathcal{A}^2 - \phi^2) = 0. \tag{2}$$

In analogy to the Ansatz used in our previous work [3] we set

$$\operatorname{grad} W = \frac{\operatorname{grad}\psi}{\frac{\partial\psi}{\partial p}}; \qquad \frac{\partial W}{\partial t} = \frac{\frac{\partial\psi}{\partial t}}{\frac{\partial\psi}{\partial p}}, \tag{3}$$

where p denotes a new parameter with the dimensions of the quantum of action. After multiplication with $(\partial\psi/\partial p)^2$ we obtain a quadratic form

$$Q = (\mathrm{grad}\psi)^2 - \frac{1}{c^2}\left(\frac{\partial\psi}{\partial t}\right)^2 - \frac{2e}{c}\frac{\partial\psi}{\partial p}\left(\mathcal{A}.\mathrm{grad}\psi + \frac{\phi}{c}\frac{\partial\psi}{\partial t}\right) + \left[\frac{c^2}{m^2} + \frac{e^2}{c^2}(\mathcal{A}^2 - \phi^2)\right]\left(\frac{\partial\psi}{\partial p}\right)^2. \quad (4)$$

We note that the coefficients of the zeroth, first and second powers of $(\partial\psi/\partial p)$ are four-dimensional invariants. Furthermore the form Q remains invariant if one sets

$$\begin{aligned} \mathcal{A} &= \mathcal{A}_1 + \mathrm{grad} f, \\ \phi &= \phi_1 - \frac{1}{c}\frac{\partial f}{\partial t}, \\ p &= p_1 - \frac{e}{c} f, \end{aligned} \quad (5)$$

where f denotes an arbitrary function of the coordinates and the time. The latter transformation leaves also the linear differential form

$$d'\Omega = \frac{e}{mc^2}(\mathcal{A}_x dx + \mathcal{A}_y dy + \mathcal{A}_z dz) - \frac{e}{mc}\phi dt + \frac{1}{mc}dp \quad (6)$$

invariant [4].

We now wish to express the form Q as the square of the gradient of the function ψ in five-dimensional space (R_5) and seek the corresponding line-element. One easily finds

$$ds^2 = dx^2 + dy^2 + dz^2 - c^2dt^2 + (d'\Omega)^2. \quad (7)$$

The Laplace equation in R_5 takes the form

$$\begin{aligned} \nabla\psi - \frac{1}{c^2}\frac{\partial^2\psi}{\partial t^2} - \frac{2e}{c}\left(\mathcal{A}.\mathrm{grad}\frac{\partial\psi}{\partial p} + \frac{\phi}{c}\frac{\partial^2\psi}{\partial t\partial p}\right) \\ - \frac{e}{c}\frac{\partial\psi}{\partial p}\left(\mathrm{div}\mathcal{A} + \frac{1}{c}\frac{\partial\phi}{\partial t}\right) \\ + \left[m^2c^2 + \frac{e^2}{c^2}(\mathcal{A}^2 - \phi^2)\right]\frac{\partial^2\psi}{\partial p^2} = 0. \end{aligned} \quad (8)$$

Like (7) and (4) it remains invariant under Lorentz transformations and the transformations [in (5)].

As the coefficients in the equation (8) do not contain the parameter p we can assume that the dependence of the wave-function ψ on p is given by an exponential factor, and in order to agree with experiment, we must then set [5]

$$\psi = \psi_o e^{2\pi i \frac{p}{h}}. \tag{9}$$

The equation for ψ_0 is invariant with respect to Lorentz transformations but not with respect to the transformations [in] (5). The significance of the superfluous parameter p seems to lie in the fact that it implements the invariance of the equations with respect to the addition of an arbitrary gradient to the four-potential.

It should be noted here that the coefficients in the equation for ψ_0 are in general complex.

If one assumes in addition that these coefficients do not depend on t and sets

$$\psi_o = e^{-\frac{2\pi i}{h}(E+mc^2)t}\psi_1, \tag{10}$$

one obtains for ψ_1 a time-independent equation which is identical with the generalization of the Schrödinger wave-equation formulated in our earlier paper. Those values of E for which there exists a function ψ_1 satisfying certain finiteness and continuity conditions are the Bohr energy levels. From the considerations just discussed it follows that the addition of a gradient to the four-potential cannot affect the energy levels. The two functions ψ_1 and $\bar{\psi}_1$, obtained with the four-potentials $\mathcal{A}$ and $\mathcal{A} - \operatorname{grad} f$, differ only by a factor $\exp(\frac{2\pi i e}{hc} f)$ of absolute value 1 and hence (under very general assumptions concerning the function f) have the same finiteness and continuity properties.

2. General Relativity

A. Wave-Equation. For the line-element in five-dimensional space we assume:

$$\begin{aligned} ds^2 &= \sum_{i,k=1}^{5} \gamma_{ik} dx_i dx_k \\ &= \sum_{i,k=1}^{5} g_{ik} dx_i dx_k + \frac{e^2}{m^2}\left(\sum_{i=1}^{5} q_i dx_i\right)^2. \end{aligned} \tag{11}$$

Here the quantities g_{ik} are the components of the Einstein metric-tensor, [and] the quantities q_i for $(i = 1, 2, 3, 4)$ [are] the components of the four-potential divided by c^2, that is

$$\sum_{i=1}^{4} q_i dx_i = \frac{1}{c^2}(\mathcal{A}_x dx + \mathcal{A}_y dy + \mathcal{A}_z dz - \phi c dt). \tag{12}$$

The quantity q_5 is a constant and x_5 is the superfluous coordinate. All the coefficients are real and independent of x_5.

The quantities g_{ik} and q_i depend only on the fields, not on the character of the point-mass; the latter is represented by the factor e^2/m^2. For brevity we wish to introduce the e/m-independent quantities

$$\frac{e}{m}q_i = a_i \qquad (i = 1, 2, 3, 4, 5), \tag{13}$$

and make the following convention: for a summation from 1 to 5 the summation sign will be written explicitly and for a summation from 1 to 4 it will be suppressed. With this notation we find

$$\left.\begin{aligned} \gamma_{ik} &= g_{ik} + a_i a_k; \qquad g_{i5} = 0 \\ \gamma &= \|\gamma_{ik}\| = a_5^2 g \end{aligned}\right\} \quad (i, k = 1, 2, 3, 4, 5) \qquad (14), (15)$$

$$\left.\begin{aligned} \gamma^{lk} &= g^{lk} \\ \gamma^{5k} &= -\frac{1}{a_5}g^{ik}a_i = -\frac{a^i}{a_5} \\ \gamma^{55} &= \frac{1}{(a_5)^2}(1 + a_i a^i) \end{aligned}\right\} \quad (i, k, l = 1, 2, 3, 4). \tag{16}$$

The wave-equation corresponding to equation (8) reads

$$\sum_{i,k=1}^{5} \frac{\partial}{\partial x_i}\left(\sqrt{-\gamma}\gamma^{ik}\frac{\partial \psi}{\partial x_k}\right) = 0, \tag{17}$$

or, written more explicitly,

$$\frac{1}{\sqrt{-g}}\frac{\partial}{\partial x_i}\left(\sqrt{-g}g^{ik}\frac{\partial \psi}{\partial x_k}\right) - \frac{2}{a_5}a^i\frac{\partial^2 \psi}{\partial x_i \partial x_5} + \frac{1}{(a_5)^2}(1 + a_i a^i)\frac{\partial^2 \psi}{\partial x_5^2} = 0. \tag{18}$$

Finally, if one introduces the function ψ_0 and the potentials q_i, this equation may be written:

$$\frac{1}{\sqrt{-g}}\frac{\partial}{\partial x_i}\left(\sqrt{-g}g^{ik}\frac{\partial \psi_o}{\partial x_k}\right) - \frac{4\pi}{h}\sqrt{-1}ecq^i\frac{\partial \psi_o}{\partial x_i} - \frac{4\pi^2 c^2}{h^2}(m^2 + e^2 q_i q^i)\psi_o = 0. \tag{19}$$

B. Equations of Motion. We now wish to express the equations of motion of a charged point-mass as those of a geodesic in R_5. For this purpose we must first compute the Christoffel symbols.

For this purpose we denote the five dimensional symbols by $\left\{ {kl \atop r} \right\}_5$ and the four-dimensional ones by $\left\{ {kl \atop r} \right\}_4$. We introduce also the covariant derivative of the four-potential:

$$A_{lk} = \frac{\partial a_l}{\partial x_k} - \left\{ {kl \atop r} \right\}_4 a_r, \tag{20}$$

and split the tensor $2A_{ik}$ into its symmetric and anti-symmetric parts:

$$\begin{aligned} B_{lk} &= A_{lk} + A_{kl} \\ M_{lk} &= A_{lk} - A_{kl} = \frac{\partial a_l}{\partial x_k} - \frac{\partial a_k}{\partial x_l}. \end{aligned} \tag{21}$$

We then have

$$\begin{aligned} \left\{ {kl \atop r} \right\}_5 &= \left\{ {kl \atop r} \right\}_4 + \frac{1}{2}(a_k g^{ir} M_{il} + a_l g^{ir} M_{ik}), \\ \left\{ {kl \atop 5} \right\}_5 &= \frac{1}{2a_5} B_{lk} - \frac{1}{2a_5}(a_k a^i M_{il} + a_l a^i M_{ik}), \\ \left\{ {k5 \atop 5} \right\}_5 &= -\frac{1}{2} a^i M_{ik}, \\ \left\{ {55 \atop r} \right\}_5 &= 0, \\ \left\{ {55 \atop 5} \right\}_5 &= 0. \end{aligned} \tag{22}$$

The equations of a geodesic in R_5 then take the form:

$$\frac{d^2 x_r}{ds^2} + \left\{ {kl \atop r} \right\}_4 \frac{dx_k}{ds}\frac{dx_l}{ds} + \frac{d'\Omega}{ds} g^{ir} M_{il} \frac{dx_l}{ds} = 0, \tag{23}$$

$$\frac{d^2 x_5}{ds^2} + \frac{1}{2a_5} B_{lk} \frac{dx_k}{ds}\frac{dx_l}{ds} - \frac{1}{a_5}\frac{d'\Omega}{ds} a^i M_{il} \frac{dx_l}{ds} = 0. \tag{24}$$

Here $d'\Omega$ denotes, as before, the linear form

$$d'\Omega = a_i dx_i + a_5 dx_5. \tag{25}$$

If one multiplies the four equations (23) with a_r, the fifth (24) with a_5, and adds, one obtains an equation that can be written in the form

$$\frac{d}{ds}\left(\frac{d'\Omega}{ds}\right) = 0. \tag{26}$$

Thus

$$\frac{d'\Omega}{ds} = \text{const.} \tag{27}$$

If one multiplies (23) with $g_{r\alpha}\frac{dx_\alpha}{ds}$ and sums over r and α, on account of the anti-symmetry of M_{ik}, one obtains

$$\frac{d}{ds}\left(g_{r\alpha}\frac{dx_r}{ds}\frac{dx_\alpha}{ds}\right) = 0, \tag{28}$$

or, on introducing the proper-time τ by the formula

$$g_{ik}dx_i dx_k = -c^2 d\tau^2, \tag{29}$$

the equation

$$\frac{d}{ds}\left(\frac{d\tau}{ds}\right)^2 = 0. \tag{30}$$

Furthermore, equations (28) and (30) are, on account of

$$ds^2 = -c^2 d\tau^2 + (d'\Omega)^2, \tag{31}$$

a consequence of (26).

From the above it follows that equation (24) is a consequence of (23): accordingly we can set it aside. If in (23) we introduce the proper-time as independent variable the fifth parameter completely drops out; suppressing also the subscript 4 on the Christoffel symbol we have:

$$\frac{d^2x_r}{d\tau^2} + \left\{ {kl \atop r} \right\}\frac{dx_k}{d\tau}\frac{dx_l}{d\tau} + \frac{d'\Omega}{d\tau}g^{ir}M_{il}\frac{dx_l}{d\tau} = 0. \tag{32}$$

The last term on the left-hand side represents the Lorentz force. In special relativity the first of these equations may be written as

$$m\frac{d}{dt}\frac{dx}{d\tau} + \frac{1}{c}\frac{d'\Omega}{d\tau}\left[\frac{e}{c}\left(\dot{z}H_y - \dot{y}H_z + \frac{\partial \mathcal{A}_x}{\partial t}\right) + e\frac{\partial\phi}{\partial x}\right] = 0. \tag{33}$$

To agree with experiment the factor outside the square-bracket must have the value 1. Hence

$$\frac{d'\Omega}{d\tau} = c, \tag{34}$$

and

$$ds^2 = 0. \tag{35}$$

The paths of the point-masses are therefore null-geodesics in the five-dimensional space.

To obtain the Hamilton-Jacobi equation we set the square of the five-dimensional gradient of a function ψ equal to zero.

$$g^{ik}\frac{\partial\psi}{\partial x_i}\frac{\partial\psi}{\partial x_k}-\frac{2}{a_5}\frac{\partial\psi}{\partial x_5}a^i\frac{\partial\psi}{\partial x_i}+(1+a_ia^i)\left(\frac{1}{a_5}\frac{\partial\psi}{\partial x_5}\right)^2=0. \tag{36}$$

If we here set

$$mca_5\frac{\frac{\partial\psi}{\partial x_i}}{\frac{\partial\psi}{\partial x_5}}=\frac{\partial W}{\partial x_i}, \tag{37}$$

and introduce the potentials q_i instead of the a_i, we obtain an equation

$$g^{ik}\frac{\partial W}{\partial x_i}\frac{\partial W}{\partial x_k}-2ecq^i\frac{\partial W}{\partial x_i}+c^2(m^2+e^2q_iq^i)=0, \tag{38}$$

which can be taken as a generalization of our equation (2) which served as the starting point.

Leningrad, Physics Institute of the University, 24 July 1926

[1] The idea for this work arose from a conversation with Prof. V. Freedericksz, whom I thank also for many valuable suggestions. Note added in proof: while this paper was still in press there arrived in Leningrad the beautiful work of Oskar Klein (*ZS.f.Phys.* **37**, 895, 1926) in which the author arrives at results which are the same in principle as those of the present paper. Because of the importance of the results their derivation by a different method (a generalization of an Ansatz used in my earlier work) may be of interest.

[2] The author has kindly given me the opportunity of reading the work in manuscript.

[3] V. Fock, On Schrödinger Wave-Mechanics ZS.f.Phys. **38**,242, (1926).

[4] The symbol d' indicates that $d'\Omega$ is not a complete differential.

[5] The appearance of the parameter p in exponential form could perhaps be connected to some relationships that were pointed out by Schrödinger in ZS.f. Phys. **12**,13, 1923.

4

The Renaissance of Weyl's Idea: EM Gauge Theory

The flaw in Weyl's theory of gravitation pointed out by Einstein appeared to be fatal, in spite of Weyl's spirited response. The situation was summarized by London in his 1927 paper, when he wrote: "One can only admire the colossal boldness which led Weyl, on the basis of a purely formal correspondence, to his gauge-geometrical interpretation of electromagnetism. In gravitation it was a physical fact, the equivalence of gravitational and inertial mass, that led Einstein to his geometry. In electromagnetic theory no such fact is known: there is no reason to think of a universal influence of the electromagnetic field on the so-called rigid measuring rods (resp. clocks). On the contrary, atomic clocks for example represent scales whose independence of their history is exhibited by the sharpness of the spectral-lines, in contrast to the non-integrable scale that Weyl assumes in the presence of a magnetic field."

But London's summary, which at first glance appears to be so critical of Weyl's theory, was actually only a preamble to a further statement in which he took a much more positive attitude. What London pointed out was that while Weyl's use of his original idea was indeed incorrect, the idea "contained a much wider range of possibilities than was actually used by its author, and . . . contained nothing less than a logical path to wave-mechanics, and from this point of view had an immediate physical meaning." As we shall see below, what he meant by this statement was that the flaw in Weyl's original theory lay not in the identification of the non-integrable factor with electromagnetism but in its application as a gravitational scale factor. The correct application was as a wave-mechanical phase factor.

FIRST INKLING

From the historical point of view, the first inkling of the correct application and interpretation of Weyl's factor came in a remarkable paper published by Schrödinger in 1922. The relevant part of this paper is reproduced here. What Schrödinger noticed was that for a large number of systems satisfying the Bohr-Sommerfeld quantization conditions, the exponent of the non-integrable Weyl factor became quantized. The only question concerned the *unit* of quantization, which Schrödinger called γ. Weyl had left the normalization of γ open as there was no way of determining it within his theory. From the fact that γ had to have the dimensions of action, Schrödinger hazarded the guess that it might be of the order of one of the two known natural units of action, namely, $\frac{e^2}{c}$ and h. The first guess led to an extremely large Weyl factor. The second guess was more interesting because it permitted the factor γ to be of order h. It also suggested that γ might be pure imaginary, indeed, that $\gamma = i\hbar$, in which case the Weyl factor would reduce to unity for Bohr-Sommerfeld orbits and the experimentally observed single-valuedness of the scale would no longer be contradicted. Thus, suggested Schrödinger, the flaw in the original Weyl theory might be removed by quantum mechanics!

Schrödinger's suggestion that the path factor might be quantized was based on five examples of Bohr-Sommerfeld systems for which this was true, but the idea can be understood more economically by means of a general argument given later by London (see the 1927 paper). London's argument was the following: Let W be the Hamilton-Jacobi function W described in Chapter 1. Then

$$W = \int \partial_\mu W dx^\mu = \int \vec{\partial} W.d\vec{x} - \int \partial_t W dt = mt + \int \vec{\partial} W.d\vec{x} + \int E dt, \quad (72)$$

where E is the relativistic energy. But

$$\int E dt = \int (V dt + \vec{p}.\vec{d}x) = \int (V dt - \vec{x}.d\vec{p}) = \int (V + \vec{x}.\vec{\partial} V) dt, \quad (73)$$

and, for the class of potentials V which are homogeneous of degree -1 (e.g., Coulomb potential) the integrand in the last term vanishes. Hence for this class of potentials we have

$$W = mt + \int \vec{\partial} W.d\vec{x}. \quad (74)$$

In that case the general relationship between A_μ and W given in (27) becomes

$$e \int A_\mu(x) dx^\mu = \int \vec{\partial} W.\vec{d}x. \quad (75)$$

This is the result obtained by Schrödinger in his five examples. So far, it is a purely classical result. Schrödinger's observation was that if one then applies the Bohr-Sommerfeld quantization condition

$$\oint \vec{\partial} W.d\vec{x} = nh, \quad (76)$$

for closed orbits, where n is an integer, equation (75) implies that

$$e^{\frac{e}{\gamma}\oint A_\mu(x)dx^\mu} = e^{\frac{2\pi\hbar n}{\gamma}}. \tag{77}$$

Hence if $\gamma = i\hbar$, the Weyl factor is unity. Of course this requires that the γ, which in Weyl's original theory was arbitrary but real, should now be pure imaginary.

Strangely enough, Schrödinger does not refer to his 1922 observation in his classic 1926 papers, but that it played a role in his invention of wave mechanics is known from a letter [1] that he wrote to London in 1927.

CONNECTION BETWEEN THE GAUGE PRINCIPLE AND WEYL'S THEORY

It was left to London to establish the relation between Weyl's non-integrable scale factor and the gauge principle as it occurs in the Hamilton-Jacobi, de Broglie and Schrödinger equations. London was aware of Schrödinger's 1922 paper, and in a 1926 letter [2] to Schrödinger, he even inquired whimsically whether "it was the same Mr. Schrödinger who had written the 1922 paper." What London pointed out was that applying the gauge principle to a free wave equation was essentially the same as multiplying the free wave function by the Weyl factor. As London's paper appeared before Schrödinger's theory was completely accepted and before the appearance of Dirac's theory, his task in making the above observation was technically difficult, and he proceeded as follows: He first considered the de Broglie theory, which he defined as the theory described by the wave function

$$\psi(x) = e^{\frac{i}{\hbar}(W - m\tau)}, \tag{78}$$

where $W(x)$ is a solution of the relativistic Hamilton-Jacobi equation with the integration constants chosen so that $\psi(x)$ is single-valued. Then, by using the definition

$$l = l_o e^{\frac{ie}{\hbar}\int A_\mu dx^\mu} \tag{79}$$

of the Weyl scale factor and the relationship between W and $\int A_\mu dx^\mu$ given in (25), he obtained

$$\frac{\psi}{l} = \frac{1}{l_o} = \text{constant}. \tag{80}$$

Eureka! London effectively says at this point: "The physical object that behaves like the Weyl scale-factor is found: it is the complex amplitude of the de Broglie wave [Der physikalische Gegenstand ist gefunden, der sich so verhält wie das Weylsche Mass: die komplexe Amplitude der de Broglieschen Welle]."

He then interrupted the discussion to refer to Schrödinger's 1922 paper and to give the more general proof of Schrödinger's result summarized above.

London then went on to consider the case of the Schrödinger theory. This was a more formidable task, and he solved it by using a 5-dimensional form of the theory. He first noted that if one writes the Schrödinger wave function in the polar form

$$\psi = \rho e^{\frac{i}{\hbar}W}, \tag{81}$$

where ρ and W are real, then the Schrödinger equation becomes two real coupled equations for these real functions, namely, the usual continuity equation for ρ and

$$\left(\frac{\partial W}{\partial x^\mu} - eA_\mu\right)\left(\frac{\partial W}{\partial x_\mu} - eA^\mu\right) - \hbar^2\frac{\nabla\rho}{\rho} + m^2 = 0 \tag{82}$$

for W. The equation for W differs from the Hamilton-Jacobi equation in the appearance of the term containing $\frac{\nabla\rho}{\rho}$, but London noticed that by making the identifications

$$\frac{\partial W}{\partial x_5} = m \quad \text{and} \quad eA^5 = m\left[1 - \sqrt{1 - \frac{\hbar^2}{m^2}\left(\frac{\nabla\rho}{\rho}\right)}\ \right], \tag{83}$$

equation (82) "could be written as a 5-dimensional (massless) Hamilton-Jacobi equation," namely,

$$\frac{dx^a}{d\tau}\frac{dx_a}{d\tau} = 0 \qquad\qquad \frac{1}{m}\frac{dx_a}{d\tau} = \left(\frac{\partial W}{\partial x^a} - eA_a\right), \tag{84}$$

where $a = 1\dots5$ and τ is the five-dimensional proper time.

Accordingly, by the same argument as was used in the de Broglie case, he obtained

$$\frac{\psi}{l} = \frac{\rho}{l_o}e^{\frac{i}{\hbar}\int(\frac{\partial W}{\partial x^a} - eA_a)dx^a} = \frac{\rho}{l_o}e^{\frac{i}{m\hbar}\int\frac{dx^a}{d\tau}\frac{dx_a}{d\tau}d\tau} = \frac{\rho}{l_o}. \tag{85}$$

Equation (85) differs from its de Broglie analogue (80) in two ways. First, l is the *five*-dimensional Weyl factor; and second, the term $\rho(x)$ on the right-hand side is not constant. However, as London pointed out, $\rho(x)$ could be absorbed in the gauge potentials by making a complex gauge transformation, in which case (85) would reduce to

$$\frac{\psi}{l^*} = \frac{1}{l_o} = \text{const.} \quad \text{where} \quad l^* = l_o e^{\frac{ie}{\hbar}\int A_a^* dx^a} = l_o e^{\frac{ie}{\hbar}\int(A_a + \frac{\hbar}{ie}\frac{\partial_a\rho}{\rho})dx^a}. \tag{86}$$

It must be admitted that, because of the 5-dimensional formalism and the use of complex potentials, London's argument in the Schrödinger case is somewhat forced. But the general message of his paper was clear: "The fault in Weyl's original theory lay not in the presence of Weyl's non-integrable scale-factor but in the fact that it was real and applied to the metric. It should be converted to a phase-factor and applied to the wave-function." In fact, London's rather cumbersome argument was not really necessary, and his proposal can be summarized by saying that in the presence of an electromagnetic field, the wave function should acquire a phase factor,

$$\psi \quad \rightarrow \quad e^{\frac{ie}{\hbar}\int A_\mu dx^\mu}\psi \tag{87}$$

Thus the Weyl scale factor, which by 1927 had been abandoned even by Weyl, acquired a new lease of life as the London phase factor.

DEBUT OF THE MINIMAL PRINCIPLE IN QUANTUM MECHANICS

The remarkable feature of London's prescription (87) is that, although it was derived from Weyl's theory, it turned out to be in complete agreement with the prescription for introducing the electromagnetic field into quantum mechanics proposed by Schrödinger in the 1926 series of papers on wave mechanics (in the next-to-last section of the fourth paper). As can be seen from the extract from that paper reproduced here, Schrödinger's approach was completely different from London's. What Schrödinger did was to generalize the relativistic electromagnetic Hamilton-Jacobi equation,

$$\left(\frac{\partial W}{\partial x^\mu} - eA_\mu(x)\right)\left(\frac{\partial W}{\partial x_\mu} - eA^\mu(x)\right) + m^2 = 0, \tag{88}$$

to the relativistic electromagnetic Klein-Gordon equation,

$$\left[\left(\partial^\mu - \frac{ie}{\hbar}A^\mu(x)\right)\left(\partial_\mu - \frac{ie}{\hbar}A_\mu(x)\right) + \frac{m^2}{\hbar^2}\right]\psi(x) = 0, \tag{89}$$

by replacing the gradient $\partial_\mu W$ by the operator $\frac{\hbar}{i}\partial_\mu$ and letting the resultant operator act on a wave function.

It is clear that (89) incorporates the wave-mechanical version of the minimal principle since it can be obtained from the free wave equation by making the substitution $\partial_\mu \rightarrow \partial_\mu - \frac{ie}{\hbar}A_\mu$ and in wave mechanics ∂_μ is proportional to the momentum operator. The equivalence of the Schrödinger and London prescriptions can easily be seen by considering the generalization of Schrödinger's prescription to arbitrary wave equations, namely,

$$P(\partial_\mu)\psi(x) = 0 \quad \rightarrow \quad P(D_\mu)\psi(x) = 0 \quad \text{where} \quad D_\mu = \partial_\mu - \frac{ie}{\hbar}A_\mu. \tag{90}$$

The corresponding London prescription is

$$P(\partial_\mu)\psi(x) = 0 \quad \rightarrow \quad P(\partial_\mu)U\psi = 0 \quad \text{where} \quad U = e^{-\frac{ie}{\hbar}\int A_\mu dx^\mu}, \tag{91}$$

and from the identity

$$U^{-1}\partial_\mu U = D_\mu, \tag{92}$$

one sees at once that the two prescriptions are equivalent. Indeed, the London and Schrödinger prescription are just the integral and differential forms of the same prescription.

It is clear that the appearance of the quantum-mechanical minimal principle in equation (89) is a direct consequence of its appearance as $\pi_\mu = \partial_\mu W$ in the classical Hamilton-Jacobi equation and of the correspondence principle $\pi_\mu \rightarrow \frac{\hbar}{i}\partial_\mu$. Thus the quantum-mechanical version of the minimal principle comes from the double substitution

$$\pi_\mu \rightarrow \pi_\mu - \frac{ie}{\hbar}A_\mu \qquad \text{and} \qquad \pi_\mu \rightarrow \frac{\hbar}{i}\partial_\mu, \tag{93}$$

where the first subsitution is the (empirically based) classical electromagnetic minimal principle and the second is the quantum-mechanical correspondence principle. It is the *combination* of these two principles that links the physical theory to the geometrical covariant derivative $D_\mu = \partial_\mu + \Gamma_\mu$.

Schrödinger himself called the transition from (88) to (89) purely formal and was interested in it mainly as a means of obtaining the correct interaction of the electron with an external magnetic field. For these reasons, and possibly because, as he emphasizes, equation (89) does not include a magnetic moment term, he advanced it only tentatively and did not place any great emphasis on the gauge principle, or even on gauge invariance. Thus, he used the minimal principle implicitly but did not attach any great significance to it. Finally, it should perhaps be mentioned that although equation (89) was first written down by Schrödinger, it is usually referred to as the Klein-Gordon [3] equation and we shall refer to it by this name.

APPLICATIONS

In spite of the fact that Schrödinger had failed to emphasize the role of the minimal principle and gauge invariance, the importance of these concepts for wave mechanics was soon realized. A first example of this dawning realization was provided by the paper of Fock reproduced here. In this paper Fock derived the Schrödinger equation (and an earlier relativistic generalization of it which he had proposed) from the Laplacian equation

$$\nabla\psi - 2e\mathcal{A}_\mu\partial_\mu\frac{\partial\psi}{\partial p} - e\frac{\partial\psi}{\partial p}\partial_\mu\mathcal{A}_\mu + \left[m^2 + e^2\mathcal{A}^2\right]\frac{\partial^2\psi}{\partial p^2} = 0, \qquad (94)$$

for a 5-dimensional space with metric

$$ds^2 = d\sigma^2 + d\omega^2 \quad \text{where} \quad d\omega = \frac{e}{m}\mathcal{A}_\mu dx^\mu + \frac{1}{m}dp, \qquad (95)$$

$d\sigma^2$ being the usual 4-dimensional Minkowski metric and p the fifth coordinate. On making the Kaluza-Klein–type Ansatz

$$\psi = \psi_o e^{2\pi i\frac{p}{h}}, \qquad (96)$$

for the wave function, where ψ_o is independent of p, Fock noted that the Laplacian was invariant with respect to the five-dimensional transformations

$$\mathcal{A}_\mu \to \mathcal{A}_\mu + \partial_\mu \qquad p \to p - ef(x) \qquad f \text{ arbitrary.} \qquad (97)$$

With the Ansatz (96), the transformations (97) are nothing but the conventional electromagnetic gauge transformations. Thus, although the 5-dimensional context tends to obscure the fact, Fock explicitly noted the invariance of the Schrödinger equation with respect to electromagnetic gauge transformations.

Fock then went on to consider the Bohr energy levels and concluded that "the addition of the gradient to the four-potential does not affect the energy levels." As mentioned earlier, the fact that he thought it worthwhile to point this out shows that the gauge invariance of physical quantities had not yet been taken for granted.

The most celebrated use of the gauge principle was in Dirac's theory [4] of the electromagnetic interaction of the spinning electron. Indeed, the importance of the Dirac equation derives almost entirely from the fact that when the free Dirac equation,

$$(\gamma^\mu\partial_\mu + m)\psi(x) = 0, \qquad (98)$$

is used as a vehicle for applying the minimal principle, i.e., when it is changed to

$$(\gamma^\mu D_\mu + m)\psi(x) = 0 \quad \text{where} \quad D_\mu = \partial_\mu - ie\mathcal{A}_\mu, \qquad (99)$$

it produces the physically observed electromagnetic interaction (including the radiative corrections, which have now been verified to an extraordinary degree of accuracy). The free Dirac equation itself is nothing more than a statement that a free electron belongs to a massive, spin one-half, unitary representation of the extended Poincare group and is not distinguished from any other statement to that effect. Thus the importance of the Dirac equation is its role as a vehicle for applying the minimal principle, and, conversely, the minimal principle finds its strongest basis in the success of the Dirac equation. The two concepts are symbiotic.

One well-known aspect of the relationship is obtained by squaring the Dirac equation (99) to obtain

$$\left(D^{\mu} D_{\mu} + \frac{e}{2m}\sigma_{\mu\nu} F^{\mu\nu} + m^{2}\right) \psi(x) = 0, \tag{100}$$

where $\sigma_{\mu\nu} = \frac{i}{4}[\gamma_{\mu}, \gamma_{\nu}]$. This is just the electromagnetic Klein-Gordon equation obtained by Schrödinger except that it includes the missing magnetic moment term.

In this connection, it may be worth emphasizing that, contrary to many statements in the literature, the appearance of the correct magnetic moment in Dirac's theory is not due primarily to the relativistic nature of the theory, but to the fact that the gauge principle is applied to a spin one-half wave equation. As we have seen, no magnetic moment term was obtained by Schrödinger for the equally relativistic Klein-Gordon equation (because it describes a spinless particle), and, on the other hand, a magnetic moment term can be derived for a non-relativistic spinning electron as follows: Since the non-relativistic spinning electron is a 2-component field, it is natural [5] to write the Schrödinger equation for it in the 2-component form

$$\frac{1}{2m}(\vec{\sigma}.\vec{\partial})^{2}\psi(x) = E\psi(x). \tag{101}$$

On using the gauge principle in (101), one obtains

$$\frac{1}{2m}(\vec{\partial} + e\vec{A}(x))^{2} + \frac{e}{2m}\vec{\sigma}.B(x)\psi(x) = (E - eV(x))\psi(x), \tag{102}$$

which is just the non-relativistic Pauli equation, which was constructed to describe the observed magnetic moment.

THE EINSTEIN-WEYL CONTROVERSY REVISITED

In the light of London's dramatic reinterpretation of the Weyl phase factor, one may ask what happens to Einstein's objection to Weyl's theory, and to Schrödinger's hopes for a quantum-mechanical resolution.

Remarkably, this question was not asked until 1980, when it was asked (and answered) by C. N. Yang [6]. Yang pointed out that in the new interpretation, Einstein's conclusion that two rods taken along different paths would have different *scales* is converted to the statement that two electrically charged rods taken along different paths would have different *phases*. Is this true? Since rods are macroscopic and are not usually electrically charged, it is more natural to ask the same question concerning electrons. So the question is whether two electrons starting with the same phase and taken along different paths will acquire different phases. But this is precisely the question that was asked independently, and without reference to Einstein's objection, by Aharonov and Bohm in [7] 1959. As is well known, the experimental answer is a decisive yes. Note that by reversing the argument, one can summarize Einstein's objection to Weyl's theory as the statement that, according to atomic spectroscopy, there is no Aharonov-Bohm effect for gravitation.

The electromagnetic gauge principle, as just formulated, also permits one to summarize Weyl's 1918 theory in a rather neat way since Weyl's 1918 connection may now be interpreted as the electromagnetically covariant version of the Christoffel connection, in the sense that Weyl's 1918 connection (42) may be written in the rather simple form

$$\Gamma^{\lambda}_{\mu\nu} = \frac{1}{2} g^{\lambda\sigma} \Big(D_{\mu} g_{\sigma\nu} + D_{\nu} g_{\sigma\mu} - D_{\sigma} g_{\mu\nu} \Big), \tag{103}$$

where $D_{\mu} = \partial_{\mu} + v_{\mu}$. Since the covariant derivative is used only for charged fields this means that, in effect, Weyl was assigning an (imaginary) electric charge to the metric tensor. From this point of view, Einstein's objection to the theory is very understandable.

TOPOLOGY

Before leaving the topic of the conversion of the scale factor to a phase factor, the topological implications should be mentioned. The implications are quite deep as they change the topology of the gauge group from non-compact to compact. To see this, one notes that under a gauge transformation

$$e^{\frac{ie}{\hbar}\int^{x} A_{\mu} dy^{\mu}} \quad \rightarrow \quad e^{\frac{ie}{\hbar}\alpha(x)} \left(e^{\frac{ie}{\hbar}\int^{x} A_{\mu} dy^{\mu}} \right). \tag{104}$$

Thus the gauge factor takes its values on the circle, in contrast to Weyl's original scale factor, which took its values on the positive part of the real line. Thus the gauge group is a $U(1)$ group at each point in space, or, in fiber-bundle language, $e^{\frac{ie}{\hbar}\alpha(x)}$ is a section of a $U(1)$ fiber-bundle. The most obvious physical manifestations of the topology of the electromagnetic gauge theories are in the Aharonov-Bohm effect mentioned in the Introduction and in the appearance of vortices in superconductors [8].

The compactness of the electromagnetic gauge group does not extend to gravitational theory. But it extends to the nuclear interactions, for which the non-abelian gauge groups are all compact. The topological properties of these groups manifest themselves in the fact that the field equations admit soliton solutions [9] such as magnetic monopoles [10] and instantons [11], which are the non-abelian counterparts of the electromagnetic vortices. More importantly, the topological properties are thought to be intimately connected with the confinement of the sub-nuclear particles (quarks and gluons).

WEYL'S REACTION

Weyl's reaction to the suggestion that the non-integrable factor be transferred from the metric to the wave function and made into a phase factor was to accept it with enthusiasm. In fact he went beyond this idea and, in the now-famous paper of 1929, which we reproduce here, he suggested that gauge invariance be used as a *principle* to obtain the electromagnetic interactions. This paper formulated the gauge principle so clearly and contained so much other valuable information that it is worth a chapter to itself.

On a Remarkable Property of the Quantum-Orbits of a Single Electron

by Erwin Schrödinger in Zurich

Zeit. f. Physik 12 (1922) 13

In Weyl's geometry [reference omitted] there appears in addition to the well-known quadratic differential form, which determines the metric at individual points, a linear differential form

$$\phi_o dx^o + \phi_1 dx^1 + \phi_2 dx^2 + \phi_3 dx^3 = \phi_i dx^i,$$

which determines the metrical connection between different points. Its geometrical meaning is that the scale of a "length" l (the square of the absolute value of a vector) does not remain invariant with respect to "congruent transfer" of the length to a neighbouring point, but undergoes the change

$$dl = -l\phi_i dx^i. \tag{1}$$

Weyl discovered that, so long as one requires that the transfer of a vector be accompanied by the congruent transfer of its length, the two of them together (the metric of the individual points + the metrical relationship) define an affine connection (i.e. a concept of the parallel transport of a vector). On congruent transfer of a length along a finite section of a world-line—e.g. by parallel transfer of a vector along such a section—the scale of the length is multipled by the factor

$$e^{-\int \phi_i dx^i}, \tag{2}$$

where the line-integral is to be taken, of course, along the world-line, and depends non-trivially on the path so long as the quantities

$$f_{ik} = \frac{\partial \phi_i}{\partial x_k} - \frac{\partial \phi_k}{\partial x_i} \tag{3}$$

do not vanish. —Physically the components of the connection are identified with the gravitational field and the f_{ik} with the electromagnetic field. If these are the correct relations—and the coordinates in a space-time region are so chosen—that, at least in some approximation, x_o denotes the time (in sec) and

$x_1x_2x_3$ the Cartesian coordinates (in cm), then the ϕ_i are, up to a universal proportionality-factor, the electromagnetic potentials in the usual sense:

$$V, \quad -\frac{1}{c}\mathcal{A}_x, \quad -\frac{1}{c}\mathcal{A}_y, \quad -\frac{1}{c}\mathcal{A}_z. \tag{4}$$

If we write this factor as $\gamma^{-1}e$ where e is the elementary quantum of electric charge in CGS units, that is to say,

$$\phi_o = \gamma^{-1}eV, \quad \phi_1 = -\gamma^{-1}\frac{e}{c}\mathcal{A}_x, \quad \phi_2 = -\gamma^{-1}\frac{e}{c}\mathcal{A}_y, \quad \phi_3 = -\gamma^{-1}\frac{e}{c}\mathcal{A}_z,$$

then, since ϕ_o has the dimension $\sec^{-1}$ and eV the dimension of "energy", γ has the dimension of action (g cm^2 sec^{-1}).—The scale-factor (2) becomes

$$e^{-\frac{e}{\gamma}\int(V dt - \mathcal{A}_x dx - \mathcal{A}_y dy - \mathcal{A}_z dz)}. \tag{5}$$

The property announced in the title and which to me seems remarkable is that the "true" quantum-conditions i.e. those that are sufficient to determine the energy and thus the spectrum, are just sufficient to make the exponent of the path-factor (5) an integer multiple of $\gamma^{-1}h$ (which is a pure number according to the above) for all approximate periods of the system. As there are some ifs and buts to be attached to this statement I shall first establish it for the individual cases for which it is valid in the simple form just stated. Only then shall I discuss the the possible meaning of the statement—concerning which—let me hasten to admit—I have not made much progress.

[Schrödinger then proceeds to give five examples for which the statement is valid, namely, (1) unperturbed Kepler orbits, (2) Zeeman effect, (3) Stark effect, (4) mixed Zeeman and Stark effects with parallel axes, and (5) relativistic mass variation. But as the examples are rather long and the London analysis discussed above gives a general summary of the situation, we shall omit the details and proceed to the discussion, which is the part of Schrödinger's paper that is most relevant from the gauge-theory point of view.]

Discussion of the Results

To summarize, we have the following situation. Were the electron in its orbit to bring along with it a "length" which remained unchanged by the transfer, then, starting from any arbitrary point on the orbit, each time that the electron returned to its approximate initial position and initial state the scale of this length would appear to be multiplied by an approximately integer power of

$$e^{\frac{h}{\gamma}} \tag{22}$$

It is difficult to believe that this result is merely an accidental mathematical consequence of the quantum conditions, and has no deeper physical meaning. The somewhat imprecise form of the approximations in which it is encountered, changes nothing in this respect; we know, in any case, that the quantum orbits are not sharply defined for two reasons: First, because of the electromagnetic radiation, which, although it certainly does not exist in its classical form, must certainly correspond to something quantum-mechanical of the same order of magnitude, since otherwise the frequencies could not be correctly deduced from the correspondence principle. Second, a lack of sharpness in the quantum orbits is caused also by the fact that in most cases the motion is periodic only in a certain approximation. (E.g. in the case of the Zeeman effect the terms quadratic in the field-strengths must be neglected in principle; and, if one takes relativistic corrections into account, the Stark effect no longer belongs to the class of strictly separable problems).

Whether the electron really brings a "length" along with it in its motion is more than questionable. It is very possible that, in the course of its motion, it continually "reestablishes" itself in the sense of Weyl. It may be that the meaning of our result is to be sought in the fact that not every re-establishment tempo is permitted to the electron, but rather that the re-establishment must take place subject to a certain dependence on the quasi-periodic orbital cyles.

One is tempted to guess what value the universal constant γ must have. There are two well-established constants with the dimensions of action, namely h and $\frac{e^2}{c}$ (though for my own part I am convinced that they are not independent). Were $\gamma \simeq \frac{e^2}{c}$ the universal factor (22) would be a very large number, of the order of e^{1000}. The other possibility $\gamma \simeq h$ suggests that the pure imaginary value

$$\gamma = \frac{h}{2\pi\sqrt{-1}}$$

might be a possibility, in which case the universal factor would be equal to unity and the scale of any accompanying distance would reproduce itself after each quasi-period.—I do not dare to judge whether this would make sense in the context of Weyl geometry.

The fact that e, h, c are not the only known universal constants should also be taken into account. If one includes the (usual) gravitational constant k and any universal mass, e.g. the electron mass, one has

$$\frac{e^2}{km^2} = \text{pure number} \simeq 10^{+40}$$

Hence

$$\frac{he^2}{km^2}$$

is a "universal quantum of action" of the order of magnitude 10^{+13} ergsec.—We should recall, however, that in this matter dimensional-considerations alone are not conclusive.

Quantization as an Eigenvalue Problem

Fourth Communication

by Erwin Schrödinger in Zurich

Annalen der Physik **81** (1926) 162

Contents: 1. Elimination of the Energy Parameter from the Oscillatory Equation. The Actual Wave-Equation. Non-Conservative Systems —2. Extension of Perturbation Theory to Perturbations that contain the Time explicitly. Dispersion Theory —3. Additions to 2: Excited Atoms, Degenerate Systems, Continuous Spectra —4. Discussion of the Resonant Case —5. Generalization for an Arbitrary Perturbation —6. Relativistic-Magnetic Generalization of the Fundamental equation. —7. On the Physical Meaning of the Scalar Field.

Relativistic-Magnetic Generalization of the Fundamental Equation

As an addition to the physical problems just mentioned, in which the *magnetic* field, which up to now was completely neglected in this series of communications, plays an important role, I should like to communicate briefly here the presumable relativistic generalization of the fundamental equation (4″), though I can do this only for the single electron and only with the greatest reservation. The latter for two reasons. First the generalization is based on purely formal analogy. Second, as mentioned already in the first communication, in the case of the Kepler problem it indeed leads formally to the Sommerfeld fine-structure formula, with even the "half-integer" azimuthal and radial quanta which are now accepted as correct; but the generalization that is necessary to produce the numerically correct splittings of the hydrogen lines that, in the Bohr picture, are given by the Goudsmit-Uhlenbeck electron spin, is still missing.

The Hamilton-Jacobi partial differential equation for the Lorentz electron can be written in the form

$$\left(\frac{1}{c}\frac{\partial W}{\partial t}+\frac{e}{c}V\right)^2-\left(\frac{\partial W}{\partial x}-\frac{e}{c}\mathcal{A}_x\right)^2-\left(\frac{\partial W}{\partial y}-\frac{e}{c}\mathcal{A}_y\right)^2 -\left(\frac{\partial W}{\partial z}-\frac{e}{c}\mathcal{A}_z\right)^2-m^2c^2=0. \tag{34}$$

Here e, m, c are the charge and mass of the electron and the speed of light respectively; V, $\mathcal{A}$ are the electromagnetic potentials of an external electromagnetic field at the position of the electron. W is the action.

From the classical (relativistic) equation (34) I now seek to derive the *wave-equation* for the electron, through the following *purely formal* procedure, which, as one easily sees, leads to equation (4″) if applied to the Hamilton equation for an ordinary massive (non-relativistic) mechanical point-particle in an arbitrary field of force. —*After* squaring I replace in (34) the

$$\textit{variables} \quad \frac{\partial W}{\partial t}, \quad \frac{\partial W}{\partial x}, \quad \frac{\partial W}{\partial y}, \quad \frac{\partial W}{\partial z}, \quad \text{by the}$$

$$\textit{operators} \quad \pm\frac{h}{2\pi i}\frac{\partial}{\partial t}, \quad \pm\frac{h}{2\pi i}\frac{\partial}{\partial x}, \quad \pm\frac{h}{2\pi i}\frac{\partial}{\partial y}, \quad \pm\frac{h}{2\pi i}\frac{\partial}{\partial z}, \tag{35}$$

respectively. I then set to zero the action of the resulting second-order linear differential operator on a wave-function ψ:

$$\Delta\psi - \frac{1}{c^2}\frac{\partial^2\psi}{\partial t^2} \mp \frac{4\pi i e}{hc}\left(\frac{V}{c}\frac{\partial\psi}{\partial t} + \mathcal{A}.\mathrm{grad}\psi\right) + \frac{4\pi^2 e^2}{h^2 c^2}\left(V^2 - \mathcal{A}^2 - \frac{m^2c^4}{e^2}\right)\psi = 0 \tag{36}$$

The symbols Δ and grad have the usual elementary three-dimensional meaning. The pair of equations (36) are the presumed relativistic-magnetic generalization of (4″) for the case of a single electron and are likewise to be understood in the sense that the complex wave-function satisfies either one or the other of the two equations.

For the hydrogen atom the Sommerfeld fine-structure formula can be obtained exactly as described in the first communication and similarly for the Zeeman effect (on neglecting the term with $\mathcal{A}^2$) and the well-known selection and polarization rules and the intensity formulae; they follow from the integral relations for the spherical functions given at the end of the third communication.

For the reasons explained in the first paragraph of this section I shall omit provisionally the detailed presentation of these calculations and in the following concluding paragraphs concentrate on the "classical" and not on the incomplete relativistic-magnetic form of the theory.

[Note: The equation (4″) referred to here is the Schrödinger equation

$$\frac{\partial\psi}{\partial t} = \frac{h}{4\pi i}\left(\nabla - \frac{8\pi^2}{h^2}V\right)\psi$$

$$\frac{\partial\bar{\psi}}{\partial t} = -\frac{h}{4\pi i}\left(\nabla - \frac{8\pi^2}{h^2}V\right)\bar{\psi}, \qquad (4'')$$

and the missing term in (36) is the spin term that was later produced by the Dirac equation.—L. O'R.]

Quantum-Mechanical Interpretation of Weyl's Theory [1]

by F. London in Stuttgart

Zeit. f. Phys. 42 (1927) 375

Chapter I. Weyl's Theory

As is known, the idea of a purely local geometry, conceived first by Riemann, has recently been completed in an exceptionally beautiful and simple manner by Weyl. One may regard Riemann's conception of space as the removal of the prejudice that the curvature conditions at one point of a space determine the curvature at all points. To make this idea of Riemann meaningful it was first necessary to assume that the measuring-rod used to determine the coefficients of the metric tensor

$$ds^2 = g_{ik}dx^i dx^k$$

at each point was a "rigid" measuring-rod.

In contrast, Weyl rightly points out that the assumption of such a rigid length-scale is in contradiction with strictly local geometry and that only the ratios of the g_{ik} at a point, and not their absolute values, may be determined. Accordingly, he assumes that a gauge-measuring-rod of length l has a variation

$$dl = l\phi_i dx^i, \tag{1}$$

under an infinitesimal translation dx^i, where the proportionality factors ϕ_i are functions of position, and thereby characteristics of the scale-proportions of the space—similar to the g_{ik}. Or, if one integrates (1):

$$l = l_o e^{\int \phi_i dx^i} \tag{2}$$

($l_o = l$ at the start of the displacement). The gauge-measure depends in general on the path (is non-integrable) and is independent only in the case that the quantities

$$f_{ik} = \frac{\partial \phi_i}{\partial x^k} - \frac{\partial \phi_k}{\partial x^i} \tag{3}$$

vanish. By definition these quantities satisfy the identity

$$\frac{\partial f_{ik}}{\partial x^l} + \frac{\partial f_{kl}}{\partial x^i} + \frac{\partial f_{li}}{\partial x^k} = 0 \qquad i \neq k \neq l, \quad i, k, l = 1, 2, 3, 4 \tag{4}$$

where the dimension of the manifold is 4. The formal agreement of these four equations with the system of Maxwell equations

$$\text{curl}\mathcal{E} + \frac{1}{c}\dot{\mathcal{H}} = 0$$

$$\text{div}\mathcal{H} = 0$$

as well as some other formal analogies led Weyl to the conclusion that, up to a constant proportionality factor, the ϕ_i are to be identified with the components Φ_i of the electromagnetic four-potential and correspondingly the f_{ik} with the electromagnetic field-strengths $\mathcal{E}$ and $\mathcal{H}$. In a logical extension of the geometrical interpretation of gravitation as the curvature of Riemannian space-time, Weyl sought to interpret the remaining physical interaction, the electromagnetic field, through the scale-relationships of the space-time, characterized by the variability of the gauge-measure. One should therefore write

$$l = l_o e^{\alpha \int \Phi_i dx_i} \qquad (\alpha = \text{proportionality-factor}). \tag{2a}$$

One can only admire the colossal boldness which led Weyl, on the basis of this purely formal correspondence alone, to his gauge-geometrical interpretation of electromagnetism: In gravitational theory it was a physical fact, the equivalence of gravitational and inertial mass, that led Einstein to his geometrical interpretation. In the theory of electromagnetism no such fact was known. There was no reason to think of a universal influence of the electromagnetic field on the so-called rigid measuring-rods (respectively clocks). On the contrary, atomic clocks, for example, represent measuring-rods, whose independence of their history is shown by the sharpness of the spectral lines, in contradiction to the non-integrable measures (2*a*) which are assumed by Weyl to accompany magnetic fields. In the face of such elementary experimental evidence, it must have been an unusually strong metaphysical conviction that prevented Weyl from abandoning the idea that Nature would have to make use of the beautiful geometrical possibility that was offered. He stuck to his conviction and evaded discussion of the above-mentioned contradictions through a rather unclear re-interpretation of the concept of "real scale," which, however, robbed his theory of its immediate physical meaning and attraction.

I do not need to discuss the abstract form of the theory any further. I would rather like to show that the original form of Weyl's theory contains a much wider range of possibilities than was used by its creator, that it contains nothing less than a logical path to wave-mechanics, and from this point of view has an immediate physical meaning.

Chapter II. The Wave-Mechanics of de Broglie and Weyl's Theory

By the theory of de Broglie I mean the still incomplete pre-quantum-mechanical theory in which the wave-function for the motion of the electron (to which we confine ourselves here)

$$\psi = e^{\frac{2\pi i}{h} W(x_i)} \qquad i = 1, 2, 3, 4 \tag{5}$$

comes from a complete solution W of the Hamilton-Jacobi partial differential equation

$$\left(\frac{\partial W}{\partial x^i} - \frac{e}{c}\Phi_i\right)\left(\frac{\partial W}{\partial x_i} - \frac{e}{c}\Phi^i\right) = -m_o^2 c^2 \tag{6}$$

where the integration constants are to be determined in the usual way so that ψ is a single-valued function of the space i.e. W is additively periodic with an integral multiple of Planck's constant as period. If one is to take seriously the radical continuum picture of matter, with the dissolution of the sharply defined electron into a field that varies continuously in space and time, as is suggested by the above de Broglie theory and even more strongly by the Schrödinger theory [2] which we will consider later, one arrives at extraordinary difficulties of principle when one investigates what sense metrical statements can have in the context of the wave-continuum. For in this oscillating, fluctuating and endlessly extended medium which has replaced the sharply defined electron one finds no unchangeable discontinuities, no rigid bodies which, as reproducible measuring rods, would allow the determination [of] a scale.

I do not by any means take the view that to talk of geometry at the atomic level one must give a practical prescription for measurement; there is no such prescription in electron theory. But if one wishes to attach a definite sense to metrical statements, it seems to me that the least that should be demanded is that there should exist some real object (as a "prototype") to which such statements could be related: the width of an electron or its distance from the nucleus, for example, even if the relationship of such a statement to a practical measurement were problematical.

But such a real object is not available in the wave-continuum. In the eternally fluid ($\pi\alpha\nu\tau\alpha\rho\epsilon\iota$) formation and dissolution of waves the principle of identity is not applicable and in the continuum there is no fixed landmark which it would be suitable to use as a reproducible measure of length. The difficulty of principle in which one is trapped here would be completely hopeless had not Weyl, in his generalization of the Riemann concept of space, already created a type of space in which precisely the non-reproducibility of the gauge-measure is a logical postulate of the radical local geometry. Whereas this theory was a superfluous nuisance from the point of view of the discontinuous electron, which was believed to provide exactly these reproducible measures of lengths, from the present point of view the situation has fundamentally changed. One is, in effect, forced to retreat to Weyl's general concept of space and to attempt to apply it to the Schrödinger continuum. In doing so there appears a simple connection.

§1. Let us assume that we already possessed a measure of length l which varies according to Weyl's prescription ($2a$) and that we move it around in the ψ-field. Let us move it around with the current-velocity of the matter, the group-velocity

$$u^i = \frac{dx^i}{d\tau} = \frac{1}{m_o}\left(\frac{\partial W}{\partial x_i} - \frac{e}{c}\Phi^i\right) \tag{7}$$

I claim that, with this reasonable prescription about the path, Weyl's scalar l is numerically identical with the de Broglie field-scalar ψ. Here there are two points that we have to make precise.

In Weyl's gauge-measure a factor α was left undetermined; for this we make the hypothesis that it is equal to $\frac{2\pi ie}{hc}$. Thus

$$l = l_o e^{\frac{2\pi i}{h}\int \frac{e}{c}\Phi_i dx^i}. \tag{2a}$$

Furthermore: we shall use not exactly the ψ of equation (5) but the five-dimensional ψ modulo a factor $e^{\frac{2\pi i}{h}m_o c^2 \tau}$ as is understood in the proposals [3] of Klein, Fock and Kudar, whereby τ is understood to be the proper time. Thus

$$\psi = e^{\frac{2\pi i}{h}(W + m_o c^2 \tau)} = e^{\frac{2\pi i}{h}\left(\int \frac{\partial W}{\partial x^i}dx^i + m_o c^2 \tau\right)}. \tag{5a}$$

This quantity ψ is to be compared with the Weyl gauge-measure ($2a$) carried along by the continuum current. One obtains

$$\frac{\psi}{l} = \frac{1}{l_o} e^{\frac{2\pi i}{h}\left(\int \left(\frac{\partial W}{\partial x^i} - \frac{e}{c}\Phi_i\right)dx^i + m_o c^2 \tau\right)},$$

where the dx^i are to be determined by the current given in (7):

$$= \frac{1}{l_o} e^{\frac{2\pi i}{h}\left(\int\left(\frac{\partial W}{\partial x^i} - \frac{e}{c}\Phi_i\right)\left(\frac{\partial W}{\partial x_i} - \frac{e}{c}\Phi^i\right)\frac{d\tau}{m_o} + m_o c^2 \tau\right)}.$$

Because of the Hamilton-Jacobi equation the integrand is equal to $-m_o c^2$ and one obtains

$$\frac{\psi}{l} = \frac{1}{l_o} e^{\frac{2\pi i}{h}.\text{const}} = \text{const.} \tag{8}$$

Thus we have found the physical object that behaves like the Weyl measure: the complex amplitude of the de Broglie wave; it therefore experiences in the electromagnetic field exactly the influence that Weyl had postulated for his gauge-measure and for which—as an empty concept of the physics of the time—he had to attribute a metaphysical existence. It is therefore the prototype of the Weyl scale. And just as in gravitational theory it is a matter of choice whether we speak of deviated light rays and masses or of their geodesic motion in a Riemannian space, equation (8) now gives us the possibility of interpreting de Broglie's matter-oscillation process and its interaction with electromagnetism as a geometrical effect, by means of a Weyl space which is filled homogeneously with matter and for which the metrical connection is non-integrable.

In the absence of an electromagnetic field the gauge-scale is constant according to (2a). One should therefore also obtain a constant value for the de Broglie wave-function if one follows it with the associated current (group) velocity (with v always $< c$). This seems to be in contradiction with one of the most fundamental [of] de Broglie's results, according to which the phases of his waves propagate with a much greater phase-velocity ($u = \frac{c^2}{v}$). However, that does not apply here because we use, not exactly the de Broglie wave-function, but its five-dimensional generalization, which is dispersion-free and for which there is therefore no difference between group and phase velocity. It is easy to convince oneself immediately that, to an observer with velocity v, the plane-wave

$$\psi = e^{-\frac{2\pi i}{h}\left(\frac{m_o c^2}{\sqrt{1-\beta^2}} t - \frac{m_o v}{\sqrt{1-\beta^2}} x - m_o c^2 \tau\right)} \qquad \left(\beta = \frac{v}{c}\right)$$

has a constant phase.

A further objection, that we are comparing a density ψ with a length l seems to me also not to present any difficulty. One actually has to compare ψ with l^{-3}, which would mean only a change in the choice of the undetermined factor α. It would be more natural to conclude from the present connection that the a priori dimension of the Weyl gauge-scale should be that of the de Broglie ψ. Such a conclusion was beyond the scope of Weyl's theory since in that theory nothing is said about the nature of l.

A more serious difficulty is presented by the complex form of the path-factor. It is here not admissible to restrict oneself to its real part. This is a reflection of the fact that the wave-function itself is intrinsically complex, or in other words represents a combination of two real state-variables, namely $\psi\bar{\psi}$ and the real part of $\frac{h}{2\pi i}\ln\psi$. In this sense it is also understood that in the variational problems of quantum mechanics ψ and $\bar{\psi}$ are to be varied independently. The meaning of the fact that every length has to be regarded as complex, and that the whole Weyl variation of length shows up as a change only of phase without any change of absolute value, is a question that I should not yet like to discuss.

§2. But there remains the objection, which we indicated above, that experiment speaks against the non-integrability of the path-factor. One sees now how this difficulty must be resolved: quantum-theory permits matter to have only a discrete range of orbits and one might guess that the distinguished orbital motions allow the gauge-scale to be transported only in such a way that, on returning to the initial point, the phase has made an integral number of orbits, so that in spite of the non-integrability of the lengths, the gauge-measure at each place is unique. In fact, one is reminded of the resonance property of de Broglie waves by means of which the old Sommerfeld-Epstein quantum conditions were re-interpreted with such important consequences. It is true that this is connected to the phase-velocity; but on account of the five-dimensional generalization of the wave-function the oscillatory phenomena are dispersion-free and our current velocity is therefore identical with the phase-velocity. Through this and the identification of the wave-function ψ with the Weyl scale-factor it would seem to be already clear [4] that the Weyl measure, if it is carried along the quantum-mechanically permitted matter-current, takes part in the resonance of the de Broglie waves and leads to a unique gauge-length at every point in spite of the non-integrability of the differential expression (2a) in the presence of electromagnetic fields. If the uniqueness of the length-scale (as a generally observed fact) had been added to the Weyl theory as an extra axiom one would necessarily have been led to the system of discrete states of motion and their de Broglie waves in the "classical" quantum-theory .

I would not like to leave this subject without pointing out that this resonance property of the Weyl scale, which appears to us here as a characteristic of wave-mechanics, was already [5] suggested as a "Remarkable Property of Quantum Orbits" by Schrödinger in 1922 and was illustrated by a number of examples, although at the time the significance was not recognized. The possibility of setting $\alpha = \frac{2\pi i e}{hc}$ was also mentioned, but it was not given priority over other choices of α. Thus already at that time Schrödinger had in his hand the characteristic wave-mechanical periodicities which he was to encounter later from such an entirely different point of view.

It is, perhaps, not superfluous if I prove the Schrödinger proposal as a theorem of "classical" quantum-mechanics, independent of the wave-mechanical considerations, as originally intended. What is claimed is: The exponent of the Weyl gauge-measure, taken around a closed spatial quantum orbit, is an integral multiple of the Planck constant:

$$\oint \frac{e}{c}\Phi_i dx^i = nh \tag{9}$$

To prove that, one uses the relation already applied in §1:

$$\int \left(\frac{\partial W}{\partial x^i} - \frac{e}{c}\Phi_i\right) dx^i = -\int m_o c^2 d\tau = -\int m_o c^2 \sqrt{1 - \left(\frac{v}{c}\right)^2}\, dt$$

On account of the quantum conditions

$$\sum_{i=1}^{3} \oint \frac{\partial W}{\partial x^i} dx^i = nh$$

one obtains from this

$$\oint \left(\frac{\partial W}{\partial x_4} dx_4 - \frac{e}{c}\Phi_i dx^i\right) = -nh - \oint m_o c^2 \sqrt{1 - \left(\frac{v}{c}\right)^2}\, dt$$

Assuming that an energy-integral exists we have

$$\frac{\partial W}{\partial x_4} dx_4 = -(E_{\text{kin}} + E_{\text{pot}})dt,$$

hence

$$-\oint \frac{e}{c}\Phi_i dx^i = -nh + \oint \left(-m_o c^2 \sqrt{1 - \left(\frac{v}{c}\right)^2} + E_{\text{kin}} + E_{\text{pot}}\right) dt.$$

Here the integrals on the right-hand side vanish on account of the relativistic generalization of the Virial theorem [6] under the assumption that the potential is homogeneous of degree -1 in the x^i, from which the required result (9) follows immediately.

One sees from this derivation that the uniqueness of the Weyl measure follows only under two assumptions. These assumptions (particularly the first) are evidently critical and cannot be fully circumvented. They guarantee certain stationary relationships in space that permit one to talk of spacially closed orbits in Minkowski space, a statement which, in general, depends on the reference frame. One would have to consider these assumptions as conditions for the possible application of the principle of identity to the space.

In most cases the orbits are not exactly periodic but only quasi-periodic. In that case it can be shown that, subject to suitable continuity conditions and in a sufficiently small neighbourhood of the initial points, the Weyl scale agrees with its initial value to within an arbitrarily small amount. This is all that is required.

That here the transport of the gauge-measure proceeds with the velocity (7) of the matter is extraordinarily satisfactory; since a transport with any other velocity would be quantum-mechanically (respectively classical-mechanically) impossible. I should prefer to postpone a more detailed justification for these relationships and their incorporation into an epistimological theory of measure, as other, fundamentally different, points of view must also be taken into account. Although we have seen how Weyl's ideas may be incorporated in the contemporary physical point of view in a way that could not have been foreseen I do not believe that one can rest content. I have here emphasized the continuity aspects of quantum mechanics in a way that does not correspond to my own persuasion. Nevertheless it seemed to me proper to pursue these ideas to their logical conclusion. In this sense the contents of the next chapter are to be regarded as provisional. I hope to return to these relationships from more general physical points of view in the near future.

Chapter III. Quantum-mechanical Re-Interpretation of Weyl's Theory

The investigations of the previous Chapters were based explicitly on the preliminary-step to quantum-mechanics called "de Broglie Theory". They are therefore wrong when they are carried over directly to the Schrödinger theory—at least in those areas where the two theories differ. One can, however, at least say that the results must remain asymptotically correct in the limit of large quantum numbers, where the two theories agree.

One may characterize the advance to the Schrödinger form of wave-mechanics as bringing the 'bundling-together' of the classical-mechanical trajectories, to which de Broglie at first attributed a wave motion only superficially according to (5), to the level of a consistent wave-continuum theory. In geometrical optics the consideration of the wave-fronts and of the individual trajectories are physically equivalent. In wave-optics on the other hand, each individual wave-ray embodied in a wave-front is influenced by its neighbours. The expression of this influence is the characteristic of the Schrödinger theory when it describes the wave-function ψ through a wave-equation instead of the Jacobi equation.

On decomposition into real and imaginary parts the Schrödinger equation reads for

$$\psi = |\psi| e^{\frac{2\pi i}{h} W} \qquad (W \quad \text{real}):$$

$$\left(\frac{h}{2\pi i}\right)^2 \frac{\Box\, |\psi|}{|\psi|} + \left(\frac{\partial W}{\partial x^i} - \frac{e}{c}\Phi_i\right)\left(\frac{\partial W}{\partial x_i} - \frac{e}{c}\Phi^i\right) + m^2c^2 = 0,$$

$$\frac{\partial}{\partial x_k}\left\{|\psi|^2 \frac{e}{m}\left(\frac{\partial W}{\partial x_k} - \frac{e}{c}\Phi_k\right)\right\} = 0 \tag{10}$$

In this representation one recognizes the contrast to the de Broglie theory in the appearance of the term $\frac{\Box\, |\psi|}{|\psi|}$. At the same time it becomes evident that there are two unknown real functions. The second equation is the continuity equation for the current, whose four components are contained in the curly brackets.

There is no question but that we must at present prefer Schrödinger's theory to de Broglie's because of its conceptual basis and because of its better agreement with experiment. We certainly cannot regard its discrepancy with the Weyl theory as a failure.

If one takes into account that discrepancies appear typically for small quantum numbers there is no doubt as to the origin of the difficulties. The Weyl theory is in its scope tailored to classical mechanics and accordingly to de Broglie theory. It is therefore not to be expected or demanded that it should suit the Schrödinger theory. The task is rather to make the modification of the now obsolete Weyl theory that leads from the de Broglie theory to the Schrödinger theory; it is the Weyl theory that must be modified to correspond to the quantum-mechanical corrections of classical laws.

One can predict the direction in which the correction to the Weyl measure will proceed. Up to now it was assumed that the four-potentials Φ_i which fully describe the electromagnetic field, alone determine the path-factor. Now the situation is changed in that the four components of the field potentials Φ_i are supplemented by a fifth, the Schrödinger ψ, which, in many respects—especially in its appearance in variational problems [7]—is analogous to the Φ_i. Matter, which from the electron-theoretical viewpoint was confined within unbreachable boundaries or within field singularities, is now spread over the whole space, and whereas in the Weyl theory one could justifiably think that in 'empty' space the measure could be influenced only by the electromagnetic field, one must now take into account that the old distinction between "impenetrable" matter and the void ($\kappa\epsilon\nu o\nu$) is removed, and that one is always in the interior of the all-pervading [8] new substance ψ.

It is therefore to be expected that as well as the exterior electromagnetic fields an interior field depending only on ψ must be taken into account. Madelung [9] has proposed a 'potential' for the interaction of this field with itself. I should like to propose as a relativistic generalization of that proposal

$$e\Phi_5 = m_o c^2 \left(1 - \sqrt{1 + \left(\frac{h}{2\pi i}\right)^2 \left(\frac{\Box|\psi|}{m_o^2 c^2 |\psi|}\right)}\right). \tag{11}$$

The word 'potential' should be used with care. Φ_5 is a relativistic scalar but does not correspond to the "scalar" potential Φ_4 which appears in relativity as the time component of a four-vector. Correspondingly, Φ_5 cannot determine the change of scale along a particular world direction. Any influence that Φ_5 will have on the gauge-measure will depend only on the absolute value of the four-dimensional displacements and not on their directions. If, accordingly, one introduces a fifth coordinate through the line element $dx_5 = c d\tau$ (τ = proper time) which is not independent of the other dx_i but is put on the same footing as them [9] through the condition

$$dx_1^2 + dx_2^2 + dx_3^2 + dx_4^2 + dx_5^2 = 0, \tag{12}$$

then one would imagine that

$$l = l_o e^{\int \frac{2\pi i}{h} \sum_1^5 \frac{e}{c} \Phi_i dx^i}, \tag{13}$$

should be the quantum-mechanical generalization of the Weyl scale.

To prove the identity of (13) with Schrödinger's wave-function we must first say along which path the generalized (13) measure is to be displaced. One will again prescribe that the transport will proceed with the current-velocity of the matter. Here one must beware of the fact that the components u^i of the four-velocity are not given by (7) although the representation of the current in the second equation in (10) suggests the seperation of the factor $e\psi\bar{\psi}$ as rest-charge-density. On account of (10) the velocity components thus obtained would not satisfy the four-velocity identity [10][11]

$$u_k u^k = \frac{dx_k}{d\tau}\frac{dx^k}{d\tau} = -c^2. \tag{12'}$$

One should rather write

$$\frac{dx_k}{dx_5} \equiv \frac{u_k}{c} = \frac{\psi\bar{\psi}}{\rho}\frac{e}{m_o c}\left(\frac{\partial W}{dx^k} - \frac{e}{c}\Phi_k\right), \tag{7a}$$

where the factor

$$\rho = e\psi\bar{\psi}\sqrt{1+\left(\frac{h}{2\pi i}\right)^2\left(\frac{\Box|\psi|}{m_o^2c^2|\psi|}\right)} = e\psi\bar{\psi}\left(1-\frac{e}{m_oc^2}\Phi_5\right), \tag{14}$$

is separated as 'rest-charge-density'.

With this prescription one obtains

$$e\Phi_5 = m_oc^2\left(1-\frac{\rho}{e\psi\bar{\psi}}\right), \tag{11a}$$

and the first Schrödinger equation reads in five-dimensional form [12]

$$\sum_1^5\left(\frac{\partial W}{dx_i}-\frac{e}{c}\Phi^i\right)\left(\frac{\partial W}{dx^i}-\frac{e}{c}\Phi_i\right) = 0, \tag{10a}$$

We now compare the length l (13) along the flow ($7a$) with the Schrödinger scalar ψ. One obtains for ψ/l

$$\frac{\psi}{l} = \frac{|\psi|}{l_o}e^{\frac{2\pi i}{h}\int\sum_1^5\left(\frac{\partial W}{\partial x_i}-\frac{e}{c}\Phi^i\right)dx_i}.$$

($7a$) gives:

$$= \frac{|\psi|}{l_o}e^{\frac{2\pi i}{h}\int\sum_1^4\frac{\psi\bar{\psi}}{\rho}\frac{e}{mc}\left(\frac{\partial W}{\partial x_i}-\frac{e}{c}\Phi^i\right)\left(\frac{\partial W}{\partial x^i}-\frac{e}{c}\Phi_i\right)dx_5+\left(\frac{\partial W}{\partial x_5}-\frac{e}{c}\Phi_5\right)dx_5}$$

($11a$) gives:

$$= \frac{|\psi|}{l_o}e^{\frac{2\pi i}{h}\int\frac{\psi\bar{\psi}}{\rho}\frac{e}{mc}\sum_1^5\left(\frac{\partial W}{\partial x_i}-\frac{e}{c}\Phi^i\right)\left(\frac{\partial W}{\partial x^i}-\frac{e}{c}\Phi_i\right)dx_5} = \frac{|\psi|}{l_o}$$

the last on account of ($10a$). One obtains therefore at first not $\psi/l = \text{constant}$ but

$$\frac{\psi}{l} = \frac{|\psi|}{l_o}, \tag{8a}$$

which is a single-valued function of position [13]. But the potentials Φ_i are only determined up to an additive gradient. If one introduces in their place the potentials

$$\Phi_k^* = \Phi_k - \frac{hc}{2\pi ie}\frac{\partial}{\partial x^k}\ln|\psi|,$$

which leave the electromagnetic field-strengths invariant, one obtains $\psi/l =$ constant.

The single-valuedness of the path-factor along the current, due to the resonance of the waves, extends naturally from the de Broglie to the Schrödinger theory, so that we do not need to add anything to the considerations of Chapter 2.

Stuttgart, Physik. Inst. d. techn. Hochschule, 27 February 1927

[1] Presented in part at the meeting of the Wurttemberg Branch of the Physical Society at Stuttgart on December 18th 1926; see also a preliminary summary report in Naturwiss. **15**, 187, 1927.[2]

[2] As is well-known there are strong arguments, presented particularly by Born and his co-workers, that the whole wave-formalism is to be interpreted in a statistical manner. Insofar as the charge-density is interpreted as a statistical weight-function, it is easy to see that the same uncertainty with respect to the applicability of the axiom of identity mentioned above will remain. But as that theory has no space-time interpretation it is of little interest for the Weyl concept of space.

[3] This concept of τ, which goes back to Kudar, Ann. d. Phys. **81**, 632, 1926, agrees throughout with the recently discussed interpretation as angular coordinate for the self-rotation of the electron (Naturwissenschaften **15**, 15, 1927). This rotation angle can be regarded as a clock carried by the electron. It transforms like the proper time.

[4] This argument is not precise but will be put right immediately.

[5] E. Schrödinger, ZS. f. Phys. **12**, 13, 1922.

[6] I am not aware of a proof of the relativistic generalization of the virial theorem in the literature and hence I will present it here. We have

$$\oint\left(-m_o c^2\sqrt{1-\left(\frac{v}{c}\right)^2}+\frac{m_o c^2}{\sqrt{1-\left(\frac{v}{c}\right)^2}}+E_{\text{pot}}\right)dt$$

$$=\oint\left(\frac{m_o v^2}{\sqrt{1-\left(\frac{v}{c}\right)^2}}+E_{\text{pot}}\right)dt=\oint\left(\sum_1^3 p_i\frac{dx^i}{dt}+E_{\text{pot}}\right)dt.$$

From this we obtain by product-integration, taking account of the periodicity conditions,

$$=\oint\left(-\sum_1^3 x^i\frac{dp_i}{dt}+E_{\text{pot}}\right)dt,$$

On account of the equations of motion $\frac{dp_i}{dt}=\frac{-\partial E_{\text{pot}}}{\partial x^i}$ and thus one obtains

$$\oint\left(\sum_1^3\left(x^i\frac{\partial E_{\text{pot}}}{\partial x^i}+E_{\text{pot}}\right)dt\right.\;.$$

Here the integrand vanishes because of Euler's homogeneous function theorem.

[7] E. Schrödinger, Ann. d. Phys. **82**, 265, 1927.

[8] Since ψ satisfies a linear differential equation. Superposition principle! Nevertheless the impenetrability property seems to find its quantum-mechanical expression in the form of the Pauli exclusion principle. (P. Ehrenfest, Naturwissenschaften, **15**, 161, 1927).

[9] E. Madelung, ZS. f. Phys. **40**, 322, 1926.

[10] The appearance of this five-dimensional quadratic form is quite logical in the sense of Weyl's requirement of gauge-invariance. Although the world-line element $d\tau$, rspectively dx_5, is a relativistic invariant it is not gauge-invariant (change to another gauge changes $d\tau$) but the vanishing of the quadratic form (12) is gauge-invariant. —It seems that the five-dimensional Ansätze of Kaluza are to understood in this sense.

[11] In the following, summation over equal indices from 1 to 4 is to be understood as before unless otherwise indicated.

[12] Hereby it is to be noted that Φ_5 for its own part is an unknown which has first to be determined. A known unresolved puzzle is why the same is not true for the potentials Φ_1, Φ_2, Φ_3, Φ_4 as one would expect. (E. Schrödinger, Ann. d. Phys. **82**, 265, 1927). $\frac{\partial W}{\partial x_5} = m_0 c$ [see (5a)].

[13] One may express this line of argument in the sense of five-dimensional geometry as follows: $\left(\frac{\partial W}{\partial x^i} - \frac{e}{c}\Phi_i\right)$ is parallel to the five-current $j_i = \frac{e}{m}\psi\bar{\psi}\left(\frac{\partial W}{\partial x^i} - \frac{e}{c}\Phi_i\right)$ and dx^i is chosen parallel to the five-current j^i. The 5-current is self-orthogonal $\left(\sum_1^5 j_i j^i = 0\right)$; hence j_i is also orthogonal to dx^i and hence $\left(\frac{\partial W}{\partial x^i} - \frac{e}{c}\Phi_i\right) dx^i = 0$. Here the fifth component of the five-current is $j_5 = \rho c$. I am indebted to Mr. A. Lande for this elegant formulation.

5

Weyl's Classic, 1929

Although not fully appreciated at the time, Weyl's 1929 paper has turned out to be one of the seminal papers of the century, both from the philosophical and from the technical point of view.

From the philosophical point of view, the paper marked the completion of his 1918 ideas. He had always been convinced that there was a close analogy between gravitation and electromagnetism and was particularly impressed by the resemblance between the derivations of charge conservation and energy-momentum conservation in the respective theories. In this paper he was able to formulate the analogies between the two theories explicitly by means of the tetrad formalism and was able to overcome the objection to his 1918 theory by adopting London's reinterpretion of the non-integrable scale factor of the metric as a non-integrable phase factor of the wave function. But whereas London's formulation was tentative Weyl's formulation was complete and went beyond all previous ideas in proposing that electromagnetism be *derived* from the gauge principle. This proposal, which is something of a luxury in the case of electromagnetism, since that theory was already well-established, has turned out to be a powerful principle for deriving the nuclear interactions and to be the common principle underlying all the known fundamental interactions.

From the technical point of view, Weyl introduced in this paper many of the concepts in field theory that are now taken for granted. This can be seen by considering the six sections of the paper in turn. In the first section, he introduced the two-component spinor theory in Minkowski space and discussed the validity of parity and time-reversal invariance P and T. In this and the following section, he developed the first systematic formulation of what is now called the tetrad, or Vierbein, formalism. The aim was not only to formulate spinor theory for curved spaces but to give a systematic derivation of the Noether conservation laws and to make the resemblance between gravitation and electromagnetism manifest. In the third section, he continued the formulation of spinor theory for curved manifolds by introducing the spin connection and constructing an invariant action. In the fourth section, he derived the energy-momentum conservation laws from invariance with respect to both general coordinate transformations and Lorentz transformations of the tetrad (Noether's theorem). He then recast gravitational theory in the tetrad formalism (section five) with a view to exhibiting the analogies between it and electromagnetism. In the final section, he came to what he considered the most fundamental part of the paper, namely, the derivation of electromagnetic theory from the gauge principle.

The debut of the paper was inauspicious. Before publishing the full paper, Weyl published a short summary in the *Proceedings of the U.S. National Academy* (while on a visit to Princeton), and this resume was immediately attacked by Pauli [1], who wrote to him as follows:

> Before me lies the April edition of the *Proc. Nat. Acad (U.S.)*. Not only does it contain an article from you under "Physics" but shows that you are now in a "Physical Laboratory": from what I hear you have even been given a chair in "Physics" in America. I admire your courage; since the conclusion is inevitable that you wish to be judged, not for successes in pure mathematics, but for your true but unhappy love for physics.

Pauli's objection was partly to the intrusion of a pure mathematician into physics, especially in view of the failure of the 1918 theory, and partly due to his well-known dislike [1] of parity violation and hence of 2-component spinors. The strength of his dislike can be judged from his dismissal of the 2-component theory in the 1933 *Handbuch* article cited in Chapter 1 and from his famous reaction to Salam's proposal to use the 2-component theory for the neutrino after parity violation had been proposed in 1955: "Give my greetings to my friend Salam and tell him to think of something better."

On reading the full version of Weyl's paper, however, Pauli's attitude softened [1], and he wrote a second letter to say:

> In contrast to the nasty things I said, the essential part of my last letter has since been overtaken, particularly by your paper in *Z. f. Physik*. For this reason I have afterward even regretted that I wrote to you. After studying your paper I believe that I have really understood what you wanted to do (this was not the case in respect of the little note in the Proc. Nat. Acad.) First let me emphasize that side of the matter concerning which I am in full agreement with you: your incorporation of spinor theory into gravitational theory. I am as dissatisfied as you are with distant parallelism and your proposal to let the tetrads rotate independently at different space-points is a true solution. Your method is valid even for the massive [Dirac] case. I thereby come to the other side of the matter, namely the unsolved difficulties of the Dirac theory (two signs of m_o) and the question of the 2-component theory. In my opinion these problems will not be solved by gravitation . . . the gravitational effects will always be much too small.

In fact, Pauli accepted "what Weyl wanted to do" to such an extent that he incorporated most of it into the famous *Handbuch* article on quantum theory quoted earlier. Indeed, he incorporated the Noether theorem and the gauge principle to such an extent that many physicists who had studied the *Handbuch* article but had not read Weyl's paper attributed the introduction of these concepts to Pauli. For example C. N. Yang [2] writes:

> I studied Pauli's Handbuch article thoroughly... and I was very much impressed with the idea that charge conservation was related to the invariance of the theory under phase changes and even more impressed with the fact that gauge-invariance determined all the electromagnetic interactions, ideas which I later found out were originally due to Weyl.

As we shall see later, by 1953 Pauli had become an ardent proponent of Weyl's point of view regarding the gauge principle.

Weyl's 1929 paper is reproduced here and the contents are displayed in the accompanying abstract. As mentioned above it consists of six separate sections and each one is worth some comment.

WEYL'S §1. TWO-COMPONENT THEORY

In the first part of section 1, Weyl introduced the concept of two-component spinors and discussed the question of P and T invariance for 2-spinors in Minkowski space.

He began by defining a 2-component spinor as a vector in the 2-dimensional (self-) representation of the group $SL(2, C)$,

$$\phi \quad \to \quad \phi' = S(l)\phi \quad \text{where} \quad S(l) = e^{\omega_{ab}\sigma_{ab}} \in SL(2, C), \tag{105}$$

and the ω's and σ's are the parameters and generators of $SL(2, C)$, respectively. He then pointed out that the bilinears

$$v_a = \phi^\dagger \sigma_a \phi \qquad \sigma_a = (1, \vec{\sigma}), \tag{106}$$

where $\vec{\sigma}$ are the Pauli matrices and the dagger denotes hermitian conjugation, transform according to the 4-vector representation of $SL(2, C)$ and that this representation of $SL(2, C)$ is isomorphic to the connected part of the Lorentz group. Thus the vectors v_a are Lorentz 4-vectors. Weyl noted that there was no bilinear Lorentz *scalar* and concluded that 2-component spinors must be massless.

The mapping of an $SL(2, C)$ spinor into a Lorentz vector exhibited in (106) shows that $SL(2, C)$ is the (double) covering group for the connected Lorentz group, and with this in mind the generators σ of $SL(2, C)$ are often written as

$$\sigma_{ij} = i\epsilon_{ijk}\sigma_k \quad \text{and} \quad \sigma_{io} = \sigma_i \quad i, j = 1, 2, 3, \tag{107}$$

the σ_i being the Pauli matrices, where the σ_{ij} and σ_{io} correspond to rotations and "pure" Lorentz transformations respectively, The complex conjugate field ϕ^* belongs to the complex conjugate representation of $SL(2, C)$. This representation is inequivalent for the full $SL(2, C)$ group but equivalent for the rotation subgroup and in order to make the rotations the same for the field and its conjugate it is usual to use not the simple complex conjugate ϕ^* but $\bar{\phi} = i\sigma_2\phi^*$.

Weyl pointed out that since the fourth component v_o of the vector v_a acquires a factor $S^\dagger S > 0$ under an $SL(2, C)$ transformation, the $SL(2, C)$ transformation excludes time-reversal *T*. He considered this to be an advantage since it takes account of the fact that nature is not time-reversable. He next pointed out that since a parity transformation P changes the signs of the pure Lorentz generators σ_{io}, and thereby converts the self-representation of $SL(2, C)$ to its *in*equivalent contragredient representation, the 2-component theory also excludes the linear implementation of parity. Since this was long before the discovery of parity violation, he considered this to be a disadvantage of the theory. Indeed, since he thought that the mass problem would eventually solved by gravitation, he attributed the need for a Dirac 4-component theory to parity rather than to mass.

It is remarkable that Weyl should even consider the possibility of time reversal and parity violation at this time. In fact, Weyl not only considered these possibilities but in the edition of his RZM book published shortly afterwards made the statement: "The problem of the proton and electron will be mixed with the symmetry properties of the quantum theory with respect to interchange of left and right, past and future, and positive and negative charge." Thus (modulo the confusion between the proton and the anti-electron that was prevalant before the experimental discovery of the positron) he not only foreshadowed the later developments in P and T violation but foreshadowed the CPT theorem. All this was at a time when, as Yang [2] put it, "[N]obody, absolutely nobody, was in any way suspecting that these symmetries were related in any manner. It was only in the 1950's that the deep connection between them was discovered.... What had prompted Weyl in 1930 to write the above passage is a mystery to me."

As Weyl spinors are less familiar than Dirac spinors, it may be useful to place them in the Dirac context. The conventional basis for the Dirac spinors is the one in which the Dirac matrix γ_o is diagonal, because γ_o serves both as parity operator and mass operator in the 4-dimensional Dirac space. However, an alternative basis is one in which the Dirac matrix γ_5 is diagonal, and this basis is very natural from the Lorentz point of view because γ_5 is Lorentz-invariant. In the γ_5 basis, the Dirac representation of the Lorentz group decomposes into two inequivalent irreducible components, labeled by $\gamma_5 = \pm 1$, respectively. In this basis, the Dirac matrices γ_μ are, of course, strictly off-diagonal because they anti-commute with γ_5. The Weyl spinors are then identified as the vectors in the 2-component irreducible representations. More precisely, they are identi fied as

$$\phi = \frac{1}{2}(1+\gamma_5)\psi \quad \text{and} \quad \xi = \frac{\sigma_2}{2i}(1-\gamma_5)\psi^*. \tag{108}$$

From this point of view, the masslessness and parity violation of the Weyl spinors are simply due to the fact that γ_5 and γ_o do not commute.

Free Weyl spinors satisfy the equation

$$\vec{\sigma}.\vec{p}\psi = \pm|p|\,\psi, \tag{109}$$

where the $\pm$ refers to the eigenvalue of γ_5. Thus for such spinors γ_5 coincides with the helicity, which is defined [3] as $h = \vec{\sigma}.\vec{p}/p$. As the eigenvalue γ_5 therefore determines whether the spin is parallel or anti-parallel to the 3-momentum, it is called the *chirality*, or handedness, and Weyl spinors are said to be *chiral*.

It was only when parity was found to be violated in 1956 that the importance of chirality came to be appreciated. The subsequent discovery that parity was not only violated, but *maximally* violated [4], in the sense that the fundamental currents had the $V-A$ structure $j_\mu = \bar{\psi}\gamma_\mu(1-\gamma_5)\psi$, was simply the discovery that these currents are chiral.

The $V-A$ discovery [5] has led to a picture of the fundamental interactions in which chiral symmetry is regarded as fundamental. The idea [6] is that the fundamental interactions are chiral in their symmetric phase and the observed non-chiral phenomenology is due to a spontaneous breakdown of this symmetry. Remnants of the original chiral symmetry are still to be observed in the $V-A$ structure of the weak currents and the chirality of the neutrinos (to within experimental error). There are good astrophysical reasons [6] for assuming that the early universe was chirally symmetric and that the spontaneous breakdown of chirality was caused by the cooling of the universe due to expansion. If this is the correct picture, then Nature in its symmetric phase prefers Weyl spinors to Dirac spinors!

It is perhaps worth noting from (108) that a Dirac spinor is equivalent to a single Weyl spinor ($\phi = \xi$) if and only if the Dirac spinor is real in the sense that $\psi^* = i\sigma_2\gamma_0\psi$. In that case the Dirac spinor is called a Majorana spinor [7] and admits a real basis. Thus a Weyl spinor is completely equivalent to a Majorana spinor, each having four real components. Interestingly enough, parity can be implemented linearly on Majorana spinors because

$$\phi(\vec{p}) \to \phi^*(-\vec{p}) \qquad \Leftrightarrow \qquad \psi(\vec{p}) \to \gamma_o\psi(-\vec{p}). \tag{110}$$

However, the Weyl version of (110) shows that it is really a CP transformation [5]. The CP transformation coincides with the P transformation for Majorana spinors because they are real and hence have $C = 1$.

WEYL'S §2. TETRAD FORMALISM

Weyl wished to incorporate his 2-component spinor theory into gravitational theory. It was already known that the use of spinors in gravitational theory presents new problems because, in contrast to the Lorentz group, the group of general coordinate transformations (diffeomorphisms) does not admit a 2-1 covering group. However, as had recently been pointed out by Wigner [8], the solution to this problem is to use local tetrads, a concept which had been introduced by Einstein [9] for a different purpose shortly before.

The simplest way to understand Wigner's argument is to note that from the uniqueness (up to conjugation) of the irreducible representation of the Dirac matrices, the only irreducible representation of the curved-space Dirac matrix equation

$$\{\gamma_\mu(x), \gamma_\nu(x)\} = 2g_{\mu\nu}(x) \tag{111}$$

is

$$\gamma_\mu(x) = e^a_\mu(x)\gamma_a, \tag{112}$$

where the γ_a are the flat-space Dirac matrices and the coefficients $e^a_\nu(x)$ satisfy the condition

$$\eta_{ab}e^a_\mu(x)e^b_\nu(x) = g_{\mu\nu}(x), \tag{113}$$

where η is a rigid Minkowski metric. But (113) is just the condition that the $e^a_\mu(x)$ form local tetrads [10]. Thus the solution of the Dirac matrix equation on curved spaces requires local tetrads.

Since the same kind of argument applies to any kind of spinors, Weyl was forced to introduce the concept of tetrads in order to incorporate his 2-component spinors into gravitational theory. Accordingly, in section 2, he proceeded to develop the tetrad formalism in a systematic manner. The formalism then turned out to be useful not only for handling spinors on curved spaces, but for other purposes, such as deriving the energy-momentum conservation laws and making the analogy between gravitation and electromagnetism manifest. Although Weyl introduced factors of i in order to work in Euclidean space, this is not necesssary, and for our brief summary, we shall use the Minkowski space formulation.

It is clear from (113) that every tetrad determines a Riemannian metric uniquely. But it is also clear that the converse cannot be true since the tetrad has 16 independent components and the metric has only 10. The 6 degrees of freedom left undetermined by the metric may be characterized precisely by the elements of a 6-parameter internal Lorentz group $\mathcal{L}$. To see this, let

$$\hat{e}^a_\mu(x) = h^a_b(x)e^b_\mu(x) \tag{114}$$

be any other tetrad satisfying (113). Then, from the definition of the tetrad we have

$$h(x)^a_c \eta_{ab} h^b_d(x) = \eta_{cd} \quad \text{or} \quad h^t \eta h = \eta, \tag{115}$$

where superscript t denotes transpose. But this is just the condition that $h(x) \in \mathcal{L}$. So the tetrads are determined by the metrics up to local $\mathcal{L}$ transformations.

If the elements $h(x) \in \mathcal{L}_c$, where $\mathcal{L}_c$ denotes the connected part of the local Lorentz group, are be written in the form

$$h(x) = e^{\omega_{ab}(x)\Sigma_{ab}}, \tag{115a}$$

where the $\omega_{ab}(x)$ are the parameters and the Σ's are the generators of $\mathcal{L}_c$, then the $\mathcal{L}_c$ transformations,

$$e^a_\mu(x) \quad \rightarrow \quad R^a_b(x) e^b_\mu(x) \quad \text{where} \quad R(x) = e^{\omega_{cd}(x)\Sigma_{cd}}, \tag{116}$$

may be thought of as a non-abelian generalization of the EM gauge transformations in Chapter 1 since the parameters are x-dependent and the transformations leave physical quantities (in this case, the components of the Riemannian metric) invariant. Thus at this point Weyl had already revealed one of the analogies between electromagnetism and gravitation.

Once the tetrad has been defined by (113) it can be used to convert any Riemannian tensor $T_{ij...k}(x)$ to an internal tensor $\Theta_{ab...c}$ according to the rule

$$\Theta_{ab...c}(x) = e^i_a(x) e^j_b(x) \ldots e^k_c(x) T_{ij...k}(x). \tag{117}$$

The quantities T transform as Riemannian tensors and internal-Lorentz scalars, and the quantities Θ transform the other way around.

Weyl then went on to define an internal version of the connection according to

$$A^b_{\mu c}(x) = e^b_\sigma \left(\nabla_\mu(\Gamma) e\right)^\sigma_c \equiv \Gamma^b_{\mu c} + J^b_{\mu c}, \tag{118}$$

where J_μ is the current

$$J^b_{\mu c} = e^b_\xi \partial_\mu e^\xi_c = \left(e^{-1} \partial_\mu e\right)^b_c. \tag{119}$$

In the last expression in (119), e^{-1} is to be understood as the matrix inverse, where e is considered to be an element of $GL(4, R)$, and from this one sees that the current J_μ lies in the $SL(4, R)$ Lie algebra. Note that in the flat-space limit $\Gamma = 0$ and $e \in \mathcal{L}$, the internal connection A_μ does not reduce to zero but to the current $e^{-1}\partial_\mu e$ for $e \in \mathcal{L}$. This shows that A_μ is not completely metrical. Equation (118) can, of course, be inverted to obtain Γ_μ in terms of A_μ.

Since it is defined by the Riemannian covariant derivative, it is clear that A_μ transforms covariantly with respect to the diffeomorphism group $\mathcal{D}$. On the other hand, it transforms non-covariantly with respect to $\mathcal{L}$,

$$A_\mu(x) \quad \rightarrow \quad A^h_\mu(x) = h^{-1}(x)A_\mu(x)h(x) + h^{-1}(x)\partial_\mu h(x)$$
$$h(x) \in \mathcal{L}, \tag{120}$$

the inhomogeneous term coming from the current-term J_μ in A_μ. Thus by replacing Γ_μ by A_μ one has replaced a connection with respect to $\mathcal{D}$ by a connection with respect to $\mathcal{L}$. So what advantage has been gained?

The biggest advantage, perhaps, is that it exhibits very explicitly the fact that gravitation is a gauge theory. Indeed, the analogy between (120) and the electromagnetic gauge transformations is evident. Although Weyl did not know it at the time, the analogy between (120) and the gauge transformations of the connections in non-abelian gauge theory is even more striking. (Compare, for example, equation (120) with equation (3) of the Yang-Mills paper in Chapter 8). Indeed, apart from the fact that the Lorentz group is non-compact, whereas the non-abelian gauge groups are compact, the two sets of transformations are identical.

A more technical advantage is that much of the differential geometry can be converted into algebra. For example, the statement that the space-time metric $g_{\mu\nu}$ is covariantly constant with respect to Γ_μ converts to the statement that the internal metric $\eta_{\mu\nu}$ is covariantly constant with respect to A_μ, and since $\eta_{\mu\nu}$ has no derivative, this is just the statement that A_μ is an element of the Lie algebra L of $\mathcal{L}$,

$$\nabla_\lambda(\Gamma)g_{\mu\nu}(x) = o \quad \Leftrightarrow \quad A^c_{\mu a}\eta_{cb} + A^c_{\mu b}\eta_{ac} = 0 \quad \Leftrightarrow \quad A_\mu \in \mathrm{L}. \tag{121}$$

Note that this implies that $A_{\mu bc}$ is anti-symmetric in the indices b, c.

Since the metric is a function of the tetrad it is possible to eliminate the metric from the definition of A_μ and thus express A_μ in terms of the tetrad alone. From (118) a short computation shows that A_μ is actually a linear function of the currents (119) with tetrad coefficients, namely, $A_{\mu[bc]} = e^a_\mu A_{a[bc]}$ where

$$2A_{a[bc]} = J_{abc} - J_{acb} + J_{bac} + J_{bca} - J_{cab} - J_{cba}. \tag{122}$$

This is the equation following (11) in Weyl's paper. It verifies that $A_{a[bc]}$ and hence $A_{\mu[bc]}$ is anti-symmetric in b and c. The generalization of (122) in the presence of torsion is simply

$$\tilde{A}_{\mu[bc]} = A_{\mu[bc]} + \theta_{\mu[bc]} \tag{123}$$

where θ is the contorsion tensor and was used by Utiyama (p. 231).

WEYL'S §3. SPINORS IN CURVED SPACE

From the group-theoretical point of view, the introduction of the local tetrads extends the group $\mathcal{D}$ of general coordinate transformations (diffeomorphisms), which lies at the heart of gravitational theory, to the semi-direct product group $\mathcal{D} \wedge \mathcal{L}$, where $\mathcal{L}$ is the group of local Lorentz transformations, and Weyl's first application of the tetrad formalism was to exploit this extension to solve the problem of incorporating 2-spinors into gravitational theory.

Accordingly (at the end of section 1) he defined 2-spinors $\phi_\alpha(x)$ in curved space to be 2-spinors with respect to $\mathcal{L}$, whose components were scalars with respect to diffeomorphisms. He concluded the section by giving their infinitesimal transformations with respect to $\mathcal{L}$, namely,

$$\delta_{\mathcal{L}}\phi^\alpha(x) = \frac{1}{2}\omega_{ab}(x)(\sigma_{ab})^\alpha_\beta\phi^\beta \qquad \alpha, \beta = 1, 2, \tag{124}$$

where the $\omega_{ab}(x)$ are the local parameters of $\mathcal{L}$ and the σ's are the the rigid generators of $SL(2, C)$ defined following (106). (This is equivalent to the second equation after (6) in Weyl's paper, with ω and σ identified as do and A, respectively, and with a, b instead of α, β). As might be expected, the definition (124) implies that

$$\psi^\dagger(x)\sigma_\mu(x)\psi(x), \tag{125}$$

which is invariant with respect to $\mathcal{L}$, is a Riemannian vector. The expression (125) is evidently the analogue for Weyl spinors of the Dirac vector $\bar{\psi}\gamma_\mu\psi$.

In section 3, Weyl went on to define the covariant derivative for spinors and construct the action for free spinors in a gravitational field. By considering parallel transport and using the fact that the covariant derivative for vectors can be written as

$$D_\mu = \partial_\mu + A^a_{\mu b}\Sigma^b_a, \quad \text{where} \quad (\Sigma_{ab})^{cd} = \frac{1}{2}\left(\delta^c_a\delta^d_b - \delta^c_b\delta^d_a\right) \tag{126}$$

are the generators of $\mathcal{L}$, he showed that the covariant derivative for the spinors must be of the similar form

$$D_\mu = \partial_\mu + A^a_{\mu b}\sigma^b_a, \tag{127}$$

where the σ^b_a are the generators of the spinor representation of $SL(2, C)$. Note that the σ's have vector labels to match the A_μ's but act on spinors. As might be expected, there is a corresponding formula for general spin, namely, $D_\mu = \partial_\mu + A^a_{\mu b}\tau^b_c$, where the τ's are the generators of the relevant representation of the internal $SL(2, C)$. The quantity $S_\mu = A^a_{\mu b}\tau^b_c$ is now called the spin connection.

Using the covariant derivative, Weyl then constructed the invariant action for the 2-spinor in a curved space, namely,

$$\mathrm{m} = \frac{1}{i}\int \psi^* \left(\sigma^\mu(x) D_\mu + D_\mu \sigma^\mu(x)\right)\psi$$

$$= \frac{1}{i}\int \psi^* \left(\sigma^\mu(x) D_\mu + \frac{1}{2}\sigma_a \frac{\partial e^a_\mu(x)}{\partial x_\mu}\right)\psi, \tag{128}$$

where $\sigma_\mu(x) \equiv e^a_\mu(x)\sigma_a$ and the σ_a are the unit and Pauli matrices of (106). The symmetrization in (128) guarantees hermiticity (up to a total divergence) and current conservation.

WEYL'S §4. THE NOETHER THEOREM

The next application of the tetrad formalism was to derive the conservation of linear and angular momentum using invariance with respect to (i) coordinate transformations and (ii) internal Lorentz transformations of the tetrad. The conservation laws for energy and momentum in flat space had, of course, been derived much earlier by Hilbert and had been derived for gravitational theory by Einstein and by Weyl himself in his 1918 paper. Furthermore, in the paper discussed in Chapter 1, Noether had given the formal prescription for obtaining the conserved current associated with *any* continuous symmetry of a Lagrangian. Thus, although Weyl did not refer to Noether, this particular result of his was actually a special case of Noether's theorem. Weyl's contribution was to make the result familiar to physicists in the context of field theory, to derive it using the tetrad formalism, and to exhibit the analogy between the energy-momentum and electromagnetic conservation laws.

As discussed in Chapter 1, the energy-momentum tensor density $T_{\mu\nu}(x)$ derived in the straightforward Noether manner is not always symmetric, although the symmetry of $T_{\mu\nu}(x)$ is necesssary for the conservation of angular momentum. One advantage of the tetrad formalism is that, as shown quite simply by Weyl, the energy-momentum tensor density $T_{\mu\nu}(x)$ derived by imposing the tetrad invariance (i) is automatically symmetric and thus the corresponding angular momentum is conserved.

Although the analogy between the energy-momentum and electric-charge conservation laws was one of the features of his theory that convinced Weyl that he was on the right track, it should be recalled that the conservation laws are derived from the *rigid*, i.e., space-time–independent, part of the group. The space-time–dependent part is used only to motivate the introduction of the gauge fields. Thus the gauge fields and the conservation laws are actually associated with complementary parts of the gauge groups, namely, the local subgroups $\tilde{G}(x)$, where $\tilde{G}(0) = 1$ and the rigid quotient groups $G(0)$, respectively.

Weyl's use of the Noether theorem was restricted to classical fields, but, as mentioned in Chapter 1, the classical conservation laws generalize to quantum field theory, either as Ward identities or as symmetries broken by anomalies, and it is fitting that one of the most interesting anomalies at the present time is the conformal anomaly, which is due to the breakdown at the quantum-mechanical level of the Weyl symmetry $g_{\mu\nu}(x) \to e^{\sigma(x)} g_{\mu\nu}(x)$, where $\sigma(x)$ is a scalar parameter.

WEYL'S §5. GRAVITATION

In the section on gravitation Weyl introduced no new physics but reformulated the usual theory in the tetrad formalism. In particular, he derived the conservation laws and expressed the Riemann tensor in the tetrad form

$$R^a_{\mu\nu b} = [D_\mu, D_\nu]^a_b = \partial_\mu A^a_{\nu b} - \partial_\nu A^a_{\mu b} + A^a_{\mu c} A^c_{\nu b} - A^a_{\nu c} A^c_{\mu b}, \qquad (129)$$

or in modern compact notation, $R = \partial \wedge A + [A, A]$. He thus exhibited the analogy between the Riemann tensor and the electromagnetic field strengths and foreshadowed the definition of field strengths in non-abelian gauge theory. He also derived the Einstein action in the tetrad form

$$\int \sqrt{-g} R = \int \sqrt{-g} \left(A_{a[bc]} A^{c[ab]} + A_c A^c \right) \quad \text{where} \quad A_c \equiv A^a_{[ac]}, \qquad (130)$$

where, as in the space-time formulation, a total divergence has been dropped on the right-hand side in order to obtain an expression which has no second derivatives. The formula (130) is particularly useful for dimensional reduction.

As mentioned earlier, it is clear from remarks made by Weyl in this and other works that he regarded the analogy between gravitation and electricity as profound, and both the conservation laws and the tetrad formulation emphasized this analogy.

WEYL'S §6. ELECTROMAGNETISM

In the final section, Weyl came to what he considered to be the crux of the paper, namely, the derivation of electromagnetism from the gauge principle.

The first step was to note that his two-component theory (and by extension, any spinorial theory) had an intrinsic abelian gauge freedom. The point was that the $SL(2, C)$ group in the $SL(2, C) \to \mathcal{L}_c$ homomorphism by which spinors are defined does not include the full group of linear transformations on spinor space. The latter group is $GL(2, C)$ and differs from $SL(2, C)$ by the central subgroup zI, where z is any complex number and I is the unit 2×2 matrix. Unitarity reduces z to a phase factor, $z = e^{i\alpha}$ for real α, but there is no way to reduce it further. Thus there is an intrinsic phase freedom in the spinor theory, namely,

$$\psi(x) \to e^{i\alpha}\psi(x), \tag{131}$$

where α is the group parameter.

The second step in Weyl's argument was to note that, just as it is natural to generalize the rigid Minkowski tetrad to a local tetrad, it is natural to generalize the phase exponent α to a local function, in which case the spinorial phase transformation (131) generalizes to

$$\psi(x) \to e^{i\alpha(x)}\psi(x). \tag{132}$$

Furthermore, since the exponent $\alpha(x)$ is independent of the tetrad, its locality should be an intrinsic property which persists even in the case of rigid tetrads. In other words, the curvature is used only to *motivate* the locality of the phase transformation and the latter may remain local even when the curvature vanishes.

The third step in Weyl's argument was to assign the known fermionic fields to representations of the phase group according to

$$\psi_e(x) \to e^{ie\alpha(x)}\psi_e(x), \tag{133}$$

where e is the electric charge, and to note that just as diffeomorphic invariance requires that $\partial_\mu \to \Delta_\mu = \partial_\mu + \Gamma_\mu(x)$, so invariance with respect to the abelian gauge group (133) requires that $\Delta_\mu \to D_\mu = \Delta_\mu + \frac{ie}{\hbar c}A_\mu(x)$, where the field $A_\mu(x)$ is a connection for the abelian group. Of course, since there is only one parameter $\alpha(x)$, the field $A_\mu(x)$ has only one component. In the flat-space limit the covariant derivative simplifies to $D_\mu = \partial_\mu + \frac{ie}{\hbar c}A_\mu$.

The electrodynamics was then obtained by applying the gauge principle, i.e., taking any Lagrangian density $L(\psi(x), \partial\psi(x))$, which is invariant with respect to the rigid subgroup of the phase group in order to conserve electric charge, and making it invariant with respect to the local phase group by changing all the derivatives to covariant ones,

$$L(\psi(x), \partial\psi(x)) \quad \to \quad L(\psi(x), D\psi(x)). \tag{134}$$

Of course, a kinetic term for the gauge field $A_\mu(x)$ itself has to be added, since otherwise the field $A(x)$ is reduced to a Lagrange multiplier and the theory is constrained. But gauge invariance, Lorentz invariance and the usual restrictions on the order and power of the derivatives restrict the kinetic term to multiples of $F_{\mu\nu}F^{\mu\nu}$, where $F_{\mu\nu}(x) = \partial_\mu A_\nu(x) - \partial_\nu A_\mu(x)$, and this term, which had already appeared in the 1918 paper, was duly added by Weyl. With this prescription, the formulation of the gauge principle was complete.

Today the second and third steps of Weyl's proposal are fully accepted. Indeed, their scope has been widened to include non-abelian gauge groups. The first step, however, is no longer valid and serves only a historical motivation for the other two. Weyl had identified the phase group in (132) as the electromagnetic gauge group because in 1929 the only known fermions (the electron and the proton) were electrically charged and the only known boson (the photon) was not. However, once uncharged fermions (such as the neutron and neutrino) and charged bosons (such as the $\pi^{\pm}$ mesons) had been discovered experimentally, this identification became untenable, and today we identify the charge associated with (131) as fermion number rather than the electric charge. Thus, as in the case of the Dirac equation, the motivation was a little weak, but the final result, guided by the intrinsic beauty of the mathematics, was profound.

Up to now we have spoken of the analogies between electromagnetism and gravitation but the gauge principle serves to emphasize also the *differences*. The most fundamental difference, of course, is that there is no equivalence principle, or universal coupling, for electromagnetism. This manifests itself in the fact that equation (132) is not sufficient but must be replaced by equation (133) in which the electric charge is introduced explicitly. Thus the way in which matter interacts with the electromagnetic field is not uniform but is determined by the charge. From the mathematical point of view, the charge is actually the *character* of the representation of the gauge group to which the matter belongs, and we shall see later that the property of the interaction being determined by the group representation is common to all gauge theories. This is particularly important in non-abelian theories, for which the representations are no longer one-dimensional. The other differences between gravitation and electromagnetism stem from the fact that whereas gravitation uses the metric, the connection and the field strength (g, Γ, R), electromagnetism uses only the connection and the field strengths (A, F). Finally, the gravitational Lagrangian density R is linear in the field strengths, but the electromagnetic Lagrangian density F^2 is quadratic.

As we have already seen in Chapter 4, the substitution $\partial \to \partial + \frac{ie}{\hbar}A$ is equivalent to the substitution

$$\psi(x) \quad \to \quad e^{\frac{ie}{\hbar}\int^{x} A_{\mu}(y)dy^{\mu}}\,\psi(x), \tag{135}$$

and thus Weyl was led once more to the non-integrable factor of his 1918 paper, but applied now to the wave function and with a factor i, as had been suggested by Schrödinger and London. Thus the wheel had turned full cycle both for Weyl's physical theory and for his philosophy. As he said himself:

> This new principle of gauge-invariance, that derives not from speculation but from experiment, shows that electromagnetism is an accompanying

phenomenon, not of the gravitational field but of the wave-function represented by ψ.

The above theory of electromagnetism was very satisfactory to Weyl from a number of points of view. First, the theory of electromagnetism was derived from a geometrical principle but the geometrical derivation was in complete accord with the empirical one. This is the meaning of the statement just quoted to the effect that the the theory "derives not from speculation but from experiment." Second, the central feature of the theory was the covariant derivative, and in this respect the theory was similar to gravitation: As he says:

> Since gauge-invariance involves an arbitrary function $\alpha(x)$ it has the character of general relativity and can only be understood in this context.

The other similarities and dissimilarities with gravitation have already been discussed.

Even Pauli was impressed by these arguments [1] and wrote:

> Here I must admit your ability in Physics. Your earlier theory with $g'_{ik} = \lambda g_{ik}$ was pure mathematics and unphysical. Einstein was justified in criticising and scolding. Now the hour of your revenge has arrived.

As already mentioned, Pauli proceeded to incorporate many of Weyl's ideas into his *Handbuch* article and by 1953 he had become an ardent proponent of the gauge principle.

Electron and Gravitation

by Hermann Weyl in Princeton, N.J.

Zeit. f. Physik 330 56 (1929)

Introduction. Relationship of General Relativity to the quantum-theoretical field equations of the spinning electron: mass, gauge-invariance, distant-parallelism. Expected modifications of the Dirac theory. —I. Two-component theory: the wave-function ψ has only two components. —§ 1. Connection between the transformation of the ψ and the transformation of a normal tetrad in four-dimensional space. Asymmetry of past and future, of left and right. —§ 2. In General Relativity the metric at a given point is determined by a normal tetrad. Components of vectors relative to the tetrad and coordinates. Covariant differentiation of ψ. —§ 3. Generally invariant form of the Dirac action, characteristic for the wave-field of matter. —§ 4. The differential conservation law of energy and momentum and the symmetry of the energy-momentum tensor as a consequence of the double-invariance (1) with respect to coordinate transformations (2) with respect to rotation of the tetrad. Momentum and angular momentum for matter. —§ 5. Einstein's classical theory of gravitation in the new analytic formulation. Gravitational Energy. —§ 6. The electromagnetic field. From the arbitrariness of the gauge-factor in ψ appears the necessity of introducing the electromagnetic potential. Gauge-invariance and charge conservation. The space-integral of charge. The introduction of mass. Discussion and rejection of another possibility in which electromagnetism appears, not as an accompanying phenomenon of matter, but of gravitation.

Introduction

In this paper I develop in detailed form a theory which embraces gravitation, electromagnetism and matter and which was sketched in the Proc. Nat. Acad., April 1929. Many authors have noticed the connection between Einstein's theory of distant-parallelism and the spin theory of the electron [1]. In spite of certain formal agreements my Ansatz is distinguished by the fact that I dispense with distant-parallelism and stick to Einstein's classical theory of gravitation. The adaption of the Pauli-Dirac theory of the spinning electron to gravitational theory promises to lead to physically fruitful results for two reasons: 1. The Dirac theory, in which the wave-field of the electron is described by four components gives twice too many energy levels; one should therefore return to the two-component Pauli theory, without sacrificing relativistic invariance. But that is prevented by the fact that the electron mass appears in the Dirac action. Mass, however, is a gravitational effect; thus there is a hope of finding a substitute in the theory of gravitation that would produce the required corrections. 2. The Dirac field-equations for ψ together with the Maxwell equations for the four potentials f_p of the electromagnetic field have an invariance property which is formally similar to the one which I called gauge-invariance in my 1918 theory of gravitation and electromagnetism; the equations remain invariant when one makes the simultaneous substitutions

$$\psi \quad \text{by} \quad e^{i\lambda}\psi \qquad \text{and} \qquad f_p \quad \text{by} \quad f_p - \frac{\partial\lambda}{\partial x_p}$$

where λ is understood to be an arbitrary function of position in four-space. Here the factor $\frac{e}{ch}$, where $-e$ is the charge of the electron, c is the speed of light and $\frac{h}{2\pi}$ is the quantum of action, has been absorbed in f_p. The connection of this "gauge-invariance" to the conservation of electric charge has also not been explored. But a fundamental difference, which is important to obtain agreement with observation, is that the exponent of the factor multiplying ψ is not real but pure imaginary. ψ now plays the role that Einstein's ds played before. It seems to me that this new principle of gauge-invariance, which follows not from speculation but from experiment, tells us that the electromagnetic field is a necessary accompanying phenomenon, not of gravitation, but of the material wave-field represented by ψ. Since gauge-invariance involves an arbitrary function λ it has the character of "general" relativity and can naturally only be understood in that context.

I prefer not to believe in distant-parallelism for a number of reasons. First my mathematical intuition objects to accepting such an artificial geometry; I find it difficult to understand the force that would keep the the local tetrads at different points and in rotated positions in a rigid relationship. There are, I believe, two important physical reasons as well. The loosening of the rigid relationship between the tetrads at different points converts the gauge-factor $e^{i\lambda}$, which remains arbitrary with respect to ψ, from a constant to an arbitrary function of space-time. In other words, only through the loosening of the rigidity does the established gauge-invariance become understandable. Secondly the freedom of rotating the tetrads independently at different points is, as we shall see, equivalent to the symmetry and conservation of the energy-momentum tensor.

In every attempt to establish the quantum-theoretical field equations one must keep in mind that these are not to be compared directly with experiment but yield only, after quantization, the basis for statistical predictions concerning the behaviour of matter and light-quanta. The Maxwell-Dirac equations contain in their present form only the electromagnetic potentials f_p and the wave-function ψ of the electron. Without doubt the wave-function ψ' of the proton must also be included. Furthermore, in the field equations ψ, ψ' and f_p should be functions of the same space-time point, before quantization one should not require for example that ψ be a function of the space-time point (t, x, y, z) and ψ' be a function of an independent space-time point (t', x', y', z'). It is reasonable to expect that in the two component-pairs of the Dirac field one pair should correspond to the electron and the other to the proton. Furthermore there should appear two electrical conservation laws, which (after quantization) should state the separate conservation of the number of electrons and protons. These would have to correspond to a two-fold gauge-invariance, involving two arbitrary functions.

We first investigate the situation in special relativity to see if, and to what extent, the formal requirements of group theory, independently of the field-equations that are to be brought into agreement with observation, force the number of components to be raised from two to four. We shall see that two components suffice if the requirement of left-right symmetry (parity) is dropped.

Two-component Theory

§1. *Transformation Law for* ψ. If one introduces the projective coordinates x_α into a space with Cartesian coordinates x, y, z:

$$x = \frac{x_1}{x_o}, \qquad y = \frac{x_2}{x_o}, \qquad z = \frac{x_3}{x_o},$$

the equation for the unit sphere becomes

$$-x_o^2 + x_1^2 + x_2^2 + x_3^2 = 0. \tag{1}$$

If one projects from the south pole to the equatorial plane $z = 0$ which is regarded as the carrier of the complex variables

$$x + iy = \zeta = \frac{\psi_2}{\psi_1}$$

the equations may be written as

$$\begin{aligned} x_o &= \bar{\psi}_1\psi_1 + \bar{\psi}_2\psi_2, & x_1 &= \bar{\psi}_1\psi_2 + \bar{\psi}_2\psi_1, \\ x_2 &= i(-\bar{\psi}_1\psi_2 + \bar{\psi}_2\psi_1), & x_3 &= i(\bar{\psi}_1\psi_1 - \bar{\psi}_2\psi_2). \end{aligned} \tag{2}$$

Thus the x_α are hermitian forms of ψ_1, ψ_2. Here, only the ratios of the variables ψ_1, ψ_2 and the coordinates x_α are relevant. A homogeneous linear transformation of ψ_1, ψ_2 (with complex coefficients) induces a real linear transformation among the coordinates x_α: it represents a collineation that leaves the unit sphere and the sense of rotation on the unit sphere invariant. It is well-known and easy to show that every such collineation can be obtained uniquely in this way.

Going from the homogeneous to the inhomogeneous, one now considers the x_α as coordinates in a four-dimensional world and (1) as the equation for the "light-cone"; and one restricts oneself to those linear transformations U of ψ_1, ψ_2 for which the determinant has absolute value 1. U induces a Lorentz transformation on the x_α, that is a real homogeneous transformation, which leaves the form

$$-x_o^2 + x_1^2 + x_2^2 + x_3^2$$

invariant. However, the formula for x_0 and our remarks about the invariance of the sense of rotation tell us immediately that the Lorentz transformations contain only a single closed continuous set Λ which (1) preserve past and future and (2) have determinant 1 and not -1; but are otherwise unconstrained. The linear transformation U is not fully determined by Λ, an arbitrary factor $e^{i\lambda}$ of absolute value 1 being left at our disposal. One may normalize it by the requirement that the determinant of U is equal to 1 but even then it remains double-valued. One would like to preserve restriction 1; it is one of the most hopeful features of the ψ-theory that it can take account of the difference between past and future. The restriction 2 removes the equivalence of left and right. It is only the fact that left-right symmetry actually appears in Nature that forces us (Part II) to introduce a second pair of ψ-components.

The hermitian conjugate of a matrix $A = \|a_{ik}\|$ will be denoted by A^*:

$$a^*_{ik} = \bar{a}_{ki}.$$

Let S_α denote the coefficient-matrix of the hermitian form of the variables ψ_1, ψ_2 which represents the coordinates x_α in (2):

$$x_\alpha = \psi^* S_\alpha \psi; \tag{3}$$

where ψ denotes the column-vector ψ_1, ψ_2. Let S_0 denote the unit matrix; we have the equations

$$S_1^2 = 1, \qquad S_2 S_3 = i S_1 \tag{4}$$

and those obtained from them by cyclic permutation of the indices. From the formal point of view it is more convenient to replace the real time-component x_0 by the imaginary ix_0. The Lorentz transformations then appear as orthogonal transformations of the four quantities

$$x(o) = ix_o, \qquad x(\alpha) = x_\alpha \qquad [\alpha = 1, 2, 3].$$

Instead of (3) one writes

$$x(\alpha) = \psi^* S(\alpha)\psi. \tag{5}$$

The transformation law of the ψ-components is such that under the influence of a transformation Λ of the coordinates $x(\alpha)$ they change so that the quantities (5) experience the transformation Λ. A quantity of this kind represents the wave-function of a matter-field, as we know from the spin-phenomenon. The $x(\alpha)$ are the coordinates in an "orthogonal tetrad" $\mathbf{e}(\alpha)$; $\mathbf{e}(1), \mathbf{e}(2), \mathbf{e}(3)$ are real space–like vectors, which constitute a left-handed Cartesian coordinate-system, $\mathbf{e}(0)/i$ is a real time–like vector in the future direction. The transformation Λ describes the transformation from such a coordinate system to another of the same kind, which will be called from now on a rotation of the tetrad. We obtain the same coefficients $c(\alpha, \beta)$ whether we express the transformation Λ through the base-vectors of the tetrad or through the coordinates

$$\mathbf{x} = \sum_\alpha x(\alpha)\mathbf{e}(\alpha) = \sum_\alpha x'(\alpha)\mathbf{e}'(\alpha)$$

$$\mathbf{e}'(\alpha) = \sum_\beta c(\alpha\beta)\mathbf{e}(\beta), \qquad x'(\alpha) = \sum_\beta c(\alpha\beta)x(\beta);$$

This follows from the orthogonal character of Λ.

For the following it is necessary to compute the infinitesimal transformation

$$d\psi = dE.\psi \tag{6}$$

that corresponds to an arbitrary infinitesimal rotation $d\Omega$:

$$dx(\alpha) = \sum_\beta do(\alpha\beta).x(\beta).$$

The $do(\alpha, \beta)$ form an anti-symmetric matrix. The transformation (6) is normalized so that the trace of dE is 0. The matrix dE depends linearly and homogeneously on the $do(\alpha, \beta)$; we therefore write

$$dE = \frac{1}{2}\sum_{\alpha\beta} do(\alpha\beta).A(\alpha\beta) = \sum do(\alpha\beta).A(\alpha\beta).$$

The last sum is to be taken only over the pairs

$$(\alpha\beta) = (01), (02), (03); \quad (23), (31), (12).$$

The dependence of $A(\alpha, \beta)$ on α and β is, of course, anti-symmetric. It should not be forgotten that the coefficients $do(\alpha, \beta)$ are pure imaginary for the first three pairs and real for the second three pairs, but are otherwise arbitrary. One finds that

$$A(23) = -\frac{1}{2i}S(1), \qquad A(01) = \frac{1}{2i}S(1) \tag{7}$$

together with two analogous equations that are obtained by cyclic permutations of the indices. To establish this one simply has to check that the infinitesimal transformations dE

$$d\psi = \frac{1}{2i}S(1)\psi \quad \text{and} \quad d\psi = \frac{1}{2}S(1)\psi$$

generate the infinitesimal rotations

$$dx(0) = 0, \quad dx(1) = 0, \quad dx(2) = -x(3), \quad dx(3) = x(2)$$

and

$$dx(0) = ix(1), \quad dx(1) = -ix(0), \quad dx(2) = 0, \quad dx(3) = 0$$

respectively.

§2. *Metric and Parallel Displacement.* We now consider General Relativity. We describe the metric at a point P by giving a local normal tetrad $\mathbf{e}(\alpha)$. Only the class of local normal tetrads—which are connected to each other by the group of rotations Λ—is determined by the metric; an individual member of the class is selected arbitrarily. The laws are invariant with respect to an arbitrary rotation of the local tetrad; the rotation of the tetrad in another point P' being independent of the rotation at P. Let $\psi_1(P)$, $\psi_2(P)$ be the components of the matter-potential at the point P relative to the chosen tetrad. A vector $\mathbf{t}$ at P can be written in the form

$$\mathbf{t} = \sum_\alpha t(\alpha)\mathbf{e}(\alpha);$$

the numbers $t(\alpha)$ being its components relative to the tetrad. For an analytic representation we need in addition a coordinate system x_p; x_p are any four continuous functions of position whose values serve to distinguish different points. The laws are therefore invariant with respect to arbitrary coordinate transformations. Let $e^p(\alpha)$ be the components of $\mathbf{e}(\alpha)$ in the coordinate-system. These four 4-component quantities describe the gravitational field. The contravariant components t^p of a vector $\mathbf{t}$ with respect to the coordinate system are connected to its components $t(\alpha)$ relative to the tetrad by the equation

$$t^p = \sum_\alpha t(\alpha)e^p(\alpha);$$

Conversely, the $t(\alpha)$ are computed from the covariant components t_p in the coordinate system according to

$$t(\alpha) = \sum_p t_p e^p(\alpha).$$

These equations control the interchange of indices. I have written the Greek indices which refer to the tetrad as arguments because for them there is no distinction between upper and lower. The interchange in the opposite sense proceeds through the matrix $\|e_p(\alpha)\|$ inverse to $\|e^p(\alpha)\|$:

$$\sum_\alpha e_p(\alpha)e^q(\alpha) = \delta_p^q \quad \text{and} \quad \sum_p e_p(\alpha)e^p(\beta) = \delta(\alpha, \beta).$$

where δ is either 0 or 1 according [to whether] the indices are the same or not. The summation rule will be used for both Latin and Greek indices. Let ϵ denote the absolute value of the determinant $|e^p(\alpha)|$. A quantity denoted by a Latin letter and divided by ϵ will, as usual, be denoted by the corresponding German letter; e.g.

$$\mathfrak{e}^p(\alpha) = \frac{e^p(\alpha)}{\epsilon}.$$

Vectors and tensors may be described by either their coordinate or tetrad components. For the quantity ψ, however, only the tetrad components make sense because the transformation of the components is according to a representation of the rotation group that cannot be extended to the group of all linear transformations. This is why, in the theory of matter, it is necessary [2] to represent the gravitational field analytically in the above form instead of the metrical form

$$\sum_{p,q} g_{pq} dx_p dx_q.$$

In addition

$$g_{pq} = e_p(\alpha)e_q(\alpha).$$

Gravitational theory must be re-cast in this new analytic form. I begin with the formula for the infinitesimal parallel transport determined by the metric. Let the vector $\mathbf{e}(\alpha)$ at the point P be transported into the vector $\mathbf{e}'(\alpha)$ at the infinitely close point P'. The $\mathbf{e}'(\alpha)$ form a normal tetrad at P' which is obtained from the local tetrad $\mathbf{e}(\alpha) = \mathbf{e}(\alpha; P')$ by an infinitesimal rotation $d\boldsymbol{\Omega}$

$$\delta\mathbf{e}(\beta) = \sum_\gamma do(\beta\gamma).\mathbf{e}(\gamma), \qquad \delta\mathbf{e}(\beta) = \mathbf{e}'(\beta) - \mathbf{e}(\beta; P'). \tag{8}$$

$d\boldsymbol{\Omega}$ is linearly dependent on the transfer PP' or its components,

$$dx_p = (dx)^p = v^p = e^p(\alpha)v(\alpha).$$

Hence we write

$$d\boldsymbol{\Omega} = \boldsymbol{\Omega}_p(dx)^p, \quad do(\beta\gamma) = o_p(\beta\gamma)(dx)^p = o(\alpha;\beta\gamma)v(\alpha). \qquad (9)$$

As is well known the parallel transfer of the vector $\mathbf{t}$ with components t^p is described by an equation

$$d\mathbf{t} = -d\boldsymbol{\Gamma}.\mathbf{t}, \quad \text{i.e.} \quad dt^p = -d\Gamma^p_r.t^r, \quad d\Gamma^p_r = \Gamma^p_{rq}(dx)^q,$$

in which the symbols Γ^p_{rq}, which are independent of $\mathbf{t}$ and dx, are symmetric in r and q. We therefore have

$$\mathbf{e}'(\beta) - \mathbf{e}(\beta) = -d\boldsymbol{\Gamma}.\mathbf{e}(\beta)$$

in addition to equation (8). Subtraction of the two differences on the left-hand-side produces the differential $d\mathbf{e}(\beta) = \mathbf{e}(\beta;\,P') - \mathbf{e}(\beta;\,P)$:

$$d\mathbf{e}^p(\beta) + d\boldsymbol{\Gamma}^p_r e^r(\beta) = -do(\beta\gamma).e^p(\gamma),$$
$$\frac{\partial e^p(\beta)}{\partial x_q}.e^q(\alpha) + \Gamma^p_{rq}e^r(\beta)e^q(\alpha) = -o(\alpha;\beta\gamma)e^p(\gamma).$$

Here one may eliminate the o to obtain the known equation for determining the $\boldsymbol{\Gamma}$ if one expresses the fact that $o(\alpha;\beta\gamma)$ is anti-symmetric with respect to β and γ. One eliminates the $\boldsymbol{\Gamma}$ and computes o by making use of the fact that Γ^p_{rq} is symmetric with respect to r and q i.e. that

$$\Gamma^p(\beta,\alpha) = \Gamma^p_{rq}e^r(\beta)e^q(\alpha)$$

is symmetric in α and β:

$$\frac{\partial e^p(\alpha)}{\partial x_q}e^q(\beta) - \frac{\partial e^p(\beta)}{\partial x_q}e^q(\alpha) = \{o(\alpha;\beta\gamma) - o(\beta;\alpha\gamma)\}e^p(\gamma). \qquad (10)$$

The left-hand side consists of the components of that "commutator-product" of the vector-fields $\mathbf{e}(\alpha)$, $\mathbf{e}(\beta)$ which is invariant with respect to coordinate transformations and plays the decisive role in the Lie theory of infinitesimal transformations; it will be denoted by $[\mathbf{e}(\alpha), \mathbf{e}(\beta)]$. Since $o(\beta; \alpha\gamma)$ is anti-symmetric in α and γ one has

$$[\mathbf{e}(\alpha), \mathbf{e}(\beta)]^p = \{o(\alpha; \beta\gamma) + o(\beta; \gamma\alpha)\}e^p(\gamma)$$

or

$$o(\alpha; \beta\gamma) + o(\beta; \gamma\alpha) = [\mathbf{e}(\alpha), \mathbf{e}(\beta)](\gamma). \tag{11}$$

If one takes the three cyclic permutations of $\alpha\beta\gamma$ in this equation and adds the resultant equations with the signs $+ - +$ one obtains

$$2o(\alpha; \beta\gamma) = [\mathbf{e}(\alpha), \mathbf{e}(\beta)](\gamma) - [\mathbf{e}(\beta), \mathbf{e}(\gamma)](\alpha) + [\mathbf{e}(\gamma), \mathbf{e}(\alpha)](\beta),$$

which shows that $o(\alpha : \beta\gamma)$ is, in fact, uniquely determined. The expression obtained fulfils all the conditions, since, as easily seen, it is anti-symmetric in β and γ.

For the following we shall need in particular the contractions

$$o(\rho, \rho\alpha) = [\mathbf{e}(\alpha), \mathbf{e}(\rho)](\rho) = \frac{\partial e^p(\alpha)}{\partial x_p} - \frac{\partial e^p(\rho)}{\partial x_q} e^q(\alpha) e_p(\rho).$$

Since

$$-\epsilon.\delta\left(\frac{1}{\epsilon}\right) = \frac{\delta\epsilon}{\epsilon} = e_q(\rho).\delta e^q(\rho)$$

we have

$$o(\rho, \rho\alpha) = \epsilon\frac{\partial \mathfrak{e}^p(\alpha)}{\partial x^p}. \tag{12}$$

§3. *Action of Matter*. With the help of parallel-transport we can compute the covariant derivative not only of the vector and tensor fields but also the ψ-field. Let $\psi_a(P), \psi_a(P')$ $[a = 1, 2]$ be the components relative to the local tetrad $\mathbf{e}(\alpha)$ at P and P' respectively. The difference $\psi_a(P') - \psi_a(P) = d\psi_a$ is the usual differential. On the other hand we parallel-transport the tetrad $\mathbf{e}(\alpha)$ from P to P': $\mathbf{e}'(\alpha)$; and let ψ'_α be the components of ψ at P' with respect to the tetrad $\mathbf{e}'(\alpha)$ itself. ψ_α and ψ'_α depend only on the choice of the tetrad $\mathbf{e}(\alpha)$ at P; they have nothing to do with the local tetrad at P'. On rotation of the tetrad at P the ψ'_α transform in exactly the same way as the ψ_α , and similarly for the differentials $\delta\psi_\alpha = \psi'_\alpha - \psi_\alpha$. They are the components of the covariant differential $\delta\psi$ of ψ. $\mathbf{e}'(\alpha)$ is obtained from the local tetrad $\mathbf{e}(\alpha) = \mathbf{e}(\alpha; P')$ at P' by the infinitesimal rotation determined in §2. The corresponding infinitesimal transformation

$$dE = \frac{1}{2} do(\beta\gamma).A(\beta\gamma)$$

changes $\psi_\alpha(P')$ into ψ'_α, i.e. $\psi' - \psi(P)$ is equal to $dE.\psi$. If one adds $d\psi = \psi(P') - \psi(P)$ one obtains

$$\delta\psi = d\psi + dE.\psi. \tag{13}$$

Everything depends linearly on the displacement PP'. We shall write

$$\delta\psi = \psi_p(dx)^p = \psi(\alpha)v(\alpha), \qquad dE = E_p(dx)^p = E(\alpha)v(\alpha).$$

We find

$$\psi_p = \left(\frac{\partial}{\partial x_p} + E_p\right)\psi \quad \text{or} \quad \psi(\alpha) = \left(e^p(\alpha)\frac{\partial}{\partial x_p} + E(\alpha)\right)\psi,$$

where

$$E(\alpha) = \frac{1}{2}o(\alpha;\beta\gamma)A(\beta\gamma).$$

If ψ' is a quantity with the same transformation law as ψ the

$$\psi^* S(\alpha)\psi'$$

are the components of a vector with respect to the local tetrad. Hence

$$v'(\alpha) = \psi^* S(\alpha)\delta\psi = \psi^* S(\alpha)\psi(\beta).v(\beta)$$

is a tetrad-independent linear transformation $v \to v'$ of the vector space at P. Its trace

$$\psi^* S(\alpha)\psi(\alpha)$$

is therefore a scalar and the equation

$$i\epsilon\mathfrak{m} = \psi^* S(\alpha)\psi(\alpha) \tag{14}$$

defines a scalar density $\mathfrak{m}$ whose integral

$$\int \mathfrak{m} dx \qquad (dx = dx_o dx_1 dx_2 dx_3)$$

can be used as an action. To obtain an explicit expression for $\mathfrak{m}$ we must compute

$$S(\alpha)E(\alpha) = \frac{1}{2}S(\alpha)A(\beta\gamma)o(\alpha;\beta\gamma). \tag{15}$$

From (7) and (4) it follows that

$$S(\beta)A(\beta\alpha) = \frac{1}{2}S(\alpha) \qquad [\alpha \neq \beta, \quad \text{no summation over} \quad \beta]$$

and

$$S(\beta)A(\gamma\delta) = \frac{1}{2}S(\alpha),$$

if $\alpha\beta\gamma\delta$ is an even permutation of the indices. The terms of the first and second kind produce as a contribution to (15) the following multiples of $S(\alpha)$

$$\frac{1}{2}o(\rho;\rho\alpha) = \frac{1}{2\epsilon}\frac{\partial \mathfrak{e}^p(\alpha)}{\partial x_p}$$

and

$$o(\beta;\gamma\delta) + o(\gamma;\delta\beta) + o(\delta;\beta\gamma) = \frac{i}{2}\phi(\alpha)$$

respectively. When $\alpha\beta\gamma\delta$ is an even permutation of 0123 we have from (11)

$$\begin{aligned} i\phi(\alpha) &= [\mathbf{e}(\beta), \mathbf{e}(\gamma)](\delta) + (\text{cycl perm of } \beta\gamma\delta) \\ &= \sum + \frac{\partial e^p(\beta)}{\partial x_q} e^q(\gamma) e_p(\delta). \end{aligned} \tag{16}$$

The sum extends over the six permutations of $\beta\gamma\delta$ with alternating sign (as well as p and q). With this notation we have

$$\mathfrak{m} = \frac{1}{i}\left(\psi^* \mathfrak{e}^p(\alpha)S(\alpha)\frac{\partial \psi}{\partial x_p} + \frac{1}{2}\frac{\partial \mathfrak{e}^p(\alpha)}{\partial x_p}\psi^* S(\alpha)\psi\right) + \frac{1}{4\epsilon}\phi(\alpha)s(\alpha). \tag{17}$$

The second part is

$$= \frac{1}{4i\epsilon}\left| e_p(\alpha), e^q(\alpha), \frac{\partial e^p(\alpha)}{\partial x_q}, s(\alpha) \right|$$

(summed over p and q); each term is a determinant of four lines, which one obtains from the given line by setting $\alpha = 0, 1, 2, 3$ successively,

$$s(\alpha) \quad \text{is} \quad = \psi^* S(\alpha)\psi \tag{18}$$

It is not the action-integral

$$\int \mathfrak{h}\, dx \tag{19}$$

itself but only its variation that is important for the equations of motion. Hence it is not necessary that $\mathfrak{h}$ be real but only that the difference $\bar{\mathfrak{h}}-\mathfrak{h}$ be a divergence. In that case we shall say that $\mathfrak{h}$ is effectively real. We must therefore check how $\mathfrak{m}$ behaves from this point of view. $e^p(\alpha)$ is real for $\alpha = 1, 2, 3$, pure imaginary for $\alpha = 0$. Hence $e^p(\alpha)S(\alpha)$ is a hermitian matrix. Similarly $\phi(\alpha)$ is real for $\alpha = 1, 2, 3$, pure imaginary for $\alpha = 0$; Hence $\phi(\alpha)S(\alpha)$ is also hermitian. Accordingly

$$\bar{\mathfrak{m}} = -\frac{1}{i}\left(\frac{\partial\psi^*}{\partial x_p}\mathfrak{S}^p\psi + \frac{1}{2}\frac{\partial \mathfrak{e}^p(\alpha)}{\partial x_p}\psi^* S(\alpha)\psi\right) + \frac{1}{4\epsilon}\phi(\alpha)s(\alpha),$$

$$i(\mathfrak{m}-\bar{\mathfrak{m}}) = \psi^*\mathfrak{S}^p\frac{\partial\psi}{\partial x_p} + \frac{\partial\psi^*}{\partial x_p}\mathfrak{S}^p\psi + \frac{\partial \mathfrak{e}^p(\alpha)}{\partial x_p}\psi^* S(\alpha)\psi$$

$$= \frac{\partial}{\partial x_p}(\psi^*\mathfrak{S}^p\psi) = \frac{\partial \mathfrak{s}^p}{\partial x_p}$$

Thus $\mathfrak{m}$ is in fact effectively real. We recover the special relativistic case when we set

$$e^o(0) = -i, \quad e^1(1) = e^2(2) = e^3(3) = 1,$$

with all other $e^p(\alpha) = 0$.

§4. *Energy*. Let (19) be the action-integral for the matter in the broad sense (matter and electric field) which are to be described by ψ and the electromagnetic potential f_p. The equations of motion are obtained from the requirement that the variation be zero

$$\delta\int \mathfrak{h}dx = 0$$

for arbitrary infinitesimal variations of ψ and f_p which vanish outside a finite space-time volume. The variation of ψ gives the field-equations for the matter in the narrower sense, the variation of the f_p the field-equations for the electromagnetic field. On subjecting the $e^p(\alpha)$, which up to now were kept invariant, to an analogous variation, one obtains, on account of these field equations, an equation

$$\delta\int \mathfrak{h}dx = \int \mathfrak{t}_p(\alpha).\delta e^p(\alpha).dx, \tag{20}$$

which defines the energy tensor density $\mathfrak{t}_p(\alpha)$.

Because of the invariance of the action the expression (20) must vanish when the variations $\delta e^p(\alpha)$ are produced by

(1) the local tetrad $\mathbf{e}(\alpha)$ undergoing an infinitesimal rotation while the coordinate-system x_p is kept fixed; or

(2) the coordinates x_p undergoing an infinitesimal transformation while the tetrad is kept fixed.

The first procedure is described by the equation

$$\delta e^p(\alpha) = o(\alpha\beta).e^p(\beta).$$

Here the $o(\alpha\beta)$ form an anti-symmetric (infinitesimal) matrix whose dependence on position is arbitrary. And the vanishing of (20) implies that

$$\mathfrak{t}(\beta\alpha) = \mathfrak{t}_p(\alpha)e^p(\beta)$$

is symmetric in α and β. The symmetry of the energy tensor is thus equivalent to the first invariance property. The symmetry is not an identity but is a consequence of the electromagnetic and matter field-equations, since for a fixed ψ-field the components will change under a rotation of the tetrad!

It is somewhat more tedious to compute the $\delta e^p(\alpha)$ that is produced by the second variation. But the computations are similar to those used in the standard analytical formulation of relativity theory [3]. Let the transformed coordinates of the point P be

$$x'_p = x_p + \delta x_p, \qquad \delta x_p = \xi^p(x).$$

Let P' denote the point that in the new coordinate-system has the same coordinates x_p that P had in the old one; it has in the old system the coordinates $x_p - \delta x_p$. The vector $\mathbf{t}$ at P will have the components

$$\frac{\partial x'_p}{\partial x_q} t^q = t_p + \frac{\partial \xi^p}{\partial x_q} t^q$$

in the new coordinate system. In particular the changes experienced by the components $e^p(\alpha)$ of the fixed vector $\mathbf{e}(\alpha)$ at the fixed point P are

$$\delta' e^p(\alpha) = \frac{\partial \xi^p}{\partial x_q} e^q(\alpha)$$

On the other hand the difference between the vector $\mathbf{e}(\alpha)$ at P' and P is given by

$$de^p(\alpha) = -\frac{\partial e^p}{\partial x_q}\xi^q.$$

Hence the variation produced by the coordinate-transformation for fixed coordinate values x_p is:

$$\delta e^p(\alpha) = \frac{\partial \xi^p}{\partial x_q} e^q(\alpha) - \frac{\partial e^p(\alpha)}{\partial x_q}\xi^q.$$

Here the ξ^p are arbitrary, apart from a function that vanishes outside a finite region of space-time. If we substitute in (20) we obtain by a partial integration

$$0 = \int \left\{ \frac{\partial \mathfrak{t}_p^q}{\partial x_q} + \mathfrak{t}_q(\alpha) \frac{\partial e^q(\alpha)}{\partial x_p} \right\} \xi^p dx.$$

The quasi-conservation law of energy and momentum follows from this in the form

$$\frac{\partial \mathfrak{t}_p^q}{\partial x_q} + \frac{\partial e^q(\alpha)}{\partial x_p} \mathfrak{t}_q(\alpha) = 0. \tag{21}$$

On account of the second term this is a true conservation equation only in the case of special relativity. In general relativity it will be so only if the energy of the gravitational field is added. In special relativity integration over the space-like surface

$$x_o = t = \text{const} \tag{22}$$

with measure $d\xi = dx_1 dx_2 dx_3$ yields the time-independent components of the momentum (J_1, J_2, J_3) and energy $(-J_0)$:

$$J_p = \int \mathfrak{t}_p^o d\xi.$$

With the help of the symmetry one finds also the divergence equations

$$\frac{\partial}{\partial x_q}(x_2 \mathfrak{t}_3^q - x_3 \mathfrak{t}_2^q) = 0, \ldots,$$
$$\frac{\partial}{\partial x_q}(x_o \mathfrak{t}_1^q + x_1 \mathfrak{t}_o^q) = 0, \ldots,$$

The three equations of the first kind show that the angular momentum (M_1, M_2, M_3)

$$M_1 = \int (x_2 \mathfrak{t}_3^o - x_3 \mathfrak{t}_2^o) d\xi, \ldots,$$

is constant in time, the equations of the second kind contain the theorem of the inertia of energy.

We compute the energy density for the matter-action $\mathfrak{m}$ constructed above; we handle the two parts which appear in the decomposition of $\mathfrak{m}$ in (17) separately. For the first part we obtain after partial integration

$$\int \delta \mathfrak{m}.dx = \int u_p(\alpha) \delta \mathfrak{e}^p(\alpha).dx$$

with

$$iu_p(\alpha)=\psi^* S(\alpha)\frac{\partial\psi}{\partial x_p}-\frac{1}{2}\frac{\partial(\psi^* S(\alpha)\psi)}{\partial x_p},$$

$$=\frac{1}{2}\left(\psi^* S(\alpha)\frac{\partial\psi}{\partial x_p}-\frac{\partial\psi^*}{\partial x_p}S(\alpha)\psi\right).$$

The part of the energy coming from this term is therefore

$$\mathfrak{t}_p(\alpha)=\mathfrak{u}_p(\alpha)-e_p(\alpha).\mathfrak{u}, \qquad \mathfrak{t}_p^q=\mathfrak{u}_p^q-\delta_p^q\mathfrak{u},$$

where $\mathfrak{u}$ is the contraction $e^p(\alpha)\mathfrak{u}_p(\alpha)$. These formulae are correct also for non-constant $e^p(\alpha)$. In the second part, however, we restrict ourselves for simplicity to special relativity. In this case

$$\int \delta\mathfrak{m}.dx=\frac{1}{4i}\int\left|e_p(\alpha), e^q(\alpha), \frac{\partial\left(\delta e^p(\alpha)\right)}{\partial x_q}, s(\alpha)\right| dx$$

$$=-\frac{1}{4i}\int\left|\delta e^q(\alpha), e_p(\alpha), e^q(\alpha), \frac{\partial s(\alpha)}{\partial x_q}\right| dx$$

$$\mathfrak{t}_p(0)=-\frac{1}{4i}\left|e_p(\alpha), e^q(\alpha), \frac{\partial s(\alpha)}{\partial x_q}\right|_{\alpha=1,2,3}.$$

From this we obtain $\mathfrak{t}_p^0$ by multiplication with $-i$; thus $\mathfrak{t}_0^0=0$ and

$$\mathfrak{t}_1^o=\frac{1}{4}\left(\frac{\partial s(3)}{\partial x_2}-\frac{\partial s(2)}{\partial x_3}\right). \tag{23}$$

We combine the two parts to determine the total energy, momentum and angular momentum. From

$$\mathfrak{t}_o^o - -\frac{1}{2i}\sum_{p=1}^{3}\left(\psi^* S^p\frac{\partial\psi}{\partial x_p}-\frac{\partial\psi^*}{\partial x_p}S^p\psi\right)$$

one obtains after carrying out a partial integration on the subtracted term,

$$-J_o=-\int \mathfrak{t}_o^o d\xi=\frac{1}{i}\int\psi^*\sum_{p=1}^{3}S^p\frac{\partial\psi}{\partial x_p}d\xi$$

This leads us to consider the operator

$$\frac{1}{i}\sum_{p=1}^{3}S^p\frac{\partial}{\partial x_p}$$

as the representative of the energy of a free particle. We also have

$$J_1 = \int \mathfrak{t}_1^o d\xi = \frac{1}{2i} \int \left(\psi^* \frac{\partial \psi}{\partial x_1} - \frac{\partial \psi^*}{\partial x_1} \psi \right) d\xi$$
$$= \frac{1}{i} \int \psi^* \frac{\partial \psi}{\partial x_1} d\xi.$$

The term (23) does not contribute to the integral. The momentum is represented by the operator

$$\frac{1}{i} \left(\frac{\partial}{\partial x_1}, \frac{\partial}{\partial x_2}, \frac{\partial}{\partial x_3} \right)$$

as it should be according to Schrödinger. From the full expression for

$$x_2 \mathbf{t}_3^o - x_3 \mathbf{t}_2^o$$

one obtains finally by suitable partial integrations

$$M_1 = \int \left\{ \frac{1}{i} \psi^* \left(x_2 \frac{\partial \psi}{\partial x_3} - x_3 \frac{\partial \psi}{\partial x_2} \right) + \frac{1}{2} s(1) \right\} . d\xi.$$

In agreement with known formulae M_1 is therefore represented by the operator

$$\frac{1}{i} \left(x_2 \frac{\partial}{\partial x_3} - x_3 \frac{\partial}{\partial x_2} \right) + \frac{1}{2} S(1)$$

Since one has incorporated spin in the theory from the beginning it must reappear here; it is, however, surprising and instructive how that happens. One sees that the fundamental assumptions of quantum-mechanics are based less on principle than one had originally supposed. They are connected to the special action $\mathfrak{m}$. On the other hand this connection confirms the necessity of $\mathfrak{m}$ in its role of matter-action. Only general relativity, which leads to a unique definition of the energy through the free variation of $e^p(\alpha)$, permits us to close the circle of quantum-mechanics as was shown above.

5. *Gravitation.* We return to the transcription of Einstein's classical theory of gravitation and determine first the curvature tensor [4]. Let the line elements d and δ connect the point P to the points P_d and P_δ. The line elements δ and d respectively are then transferred to P_d and P_δ along arbitrary paths so that they meet at the corner P^* opposite to P of an infinitesimal 'parallelogram'. Let the tetrad $\mathbf{e}(\alpha)$ be parallel-transferred along the paths $P P_d P^*$ and $P P_\delta P^*$

respectively. The two normal tetrads that one obtains in this way at P^* can be obtained from another by an infinitesimal rotation

$$\mathbf{P}_{pq}(dx)^p(\delta x)^q = \frac{1}{2}\mathbf{P}_{pq}(\Delta x)^{pq}$$

where

$$(\Delta x)^{pq} = (dx)^p(\delta x)^q - (\delta x)^p(dx)^q$$

are the components of the surface spanned by dx and δx and P_{pq} is anti-symmetric with respect to p and q. P_{pq} is an anti-symmetric matrix $\|r_{pq}(\alpha\beta)\|$; it is the Riemann curvature tensor.

The rotation that produces the tetrad $\mathbf{e}^*(\alpha)$ from the local tetrad $\mathbf{e}(\alpha)$ after parallel-transfer to P^* along the first path is, in easily understood notation,

$$(1 + d\mathbf{\Omega})(1 + \delta\mathbf{\Omega}(P_d)).$$

The difference between this expression and the one obtained by interchange of d and δ is

$$= (d(\delta\mathbf{\Omega}) - \delta(d\mathbf{\Omega})) + (d\mathbf{\Omega}\delta\mathbf{\Omega} - \delta\mathbf{\Omega}d\mathbf{\Omega}).$$

$$d\mathbf{\Omega} = \mathbf{\Omega}_p(dx)^p,$$

$$\delta(d\mathbf{\Omega}) = \frac{\partial\mathbf{\Omega}_p}{\partial x_q}\delta x_q dx_p + \mathbf{\Omega}_p \delta dx_p.$$

Since the parallelogram closes we have $\delta dx_p = d\delta x_p$; Hence finally

$$\mathbf{P}_{pq} = \left(\frac{\partial\mathbf{\Omega}_p}{\partial x_q} - \frac{\partial\mathbf{\Omega}_q}{\partial x_p}\right) + \left(\mathbf{\Omega}_p\mathbf{\Omega}_q - \mathbf{\Omega}_q\mathbf{\Omega}_p\right).$$

For the scalar curvature

$$r = e^p(\alpha)e^q(\beta)r_{pq}(\alpha\beta)$$

the first, differentiated, part gives the contribution

$$\left(e^q(\alpha)e^q(\beta) - e^q(\beta)e^p(\alpha)\right)\frac{\partial o_p(\alpha\beta)}{\partial x_q}.$$

For $\mathfrak{r} = \frac{r}{\epsilon}$ it gives, after dropping a total divergence, the two terms

$$-2o(\beta;\alpha\beta)\frac{\partial \mathfrak{e}^p(\beta)}{\partial x_q},$$

and

$$\frac{1}{\epsilon} o_p(\alpha\beta) \left\{ \frac{\partial e^p(\alpha)}{\partial x_q} e^q(\beta) - \frac{\partial e^p(\beta)}{\partial x_q} e^q(\alpha) \right\}.$$

The first is, according to (12)

$$= -2o(\beta;\rho\beta)o(\alpha;\alpha\rho).$$

The second, according to (10)

$$= 2o(\alpha;\beta\gamma).o(\gamma;\alpha\beta).$$

The result is the following expression for the gravitational action-density $\mathfrak{g}$

$$\epsilon\mathfrak{g} = o(\alpha;\beta\gamma)o(\gamma;\alpha\beta) + o(\alpha;\alpha\gamma)o(\beta;\beta\gamma). \tag{24}$$

The integral $\int \mathfrak{g}dx$ is not truly invariant, but is effectively invariant as $\mathfrak{g}$ differs from the scalar density $\mathfrak{r}$ by a divergence.

Variation of the total action integral

$$\int (\mathfrak{g} + \kappa\mathfrak{h})dx$$

yields the gravitational equations (κ being a numerical constant).

One obtains from $\mathfrak{g}$ the gravitational energy $\mathfrak{v}_p^q$ by considering an infinitesimal displacement in coordinate space [5]

$$x'_p = x_p + \xi^p, \qquad \xi^p = \text{const.}$$

The resultant variation

$$\delta\mathbf{e}(\alpha) \quad \text{is} \quad = -\frac{\partial\mathbf{e}(\alpha)}{\partial x_p}\xi^p.$$

$\mathfrak{g}$ is a function of $e^p(\alpha)$ and the derivatives $e_q^p(\alpha) = \frac{\partial e^p(\alpha)}{\partial x_q}$; the total derivative will be denoted by

$$\delta\mathfrak{g} = \mathfrak{g}_p(\alpha)\delta e^p(\alpha) + \mathfrak{g}_p^q(\alpha)\delta e_q^p(\alpha).$$

For the variations due to the infinitesimal translations in coordinate space we must have

$$\int \delta\mathfrak{g}dx + \int \frac{\partial\mathfrak{g}}{\partial x_p}\xi^p dx = 0, \tag{25}$$

the integral being taken over an arbitrary volume of space-time,

$$\int \delta \mathfrak{g} dx = \int \left(\mathfrak{g}(\alpha) - \frac{\partial \mathfrak{g}_p^q(\alpha)}{\partial x_q} \right) \delta e^p(\alpha) dx + \int \frac{\partial \left(\mathfrak{g}_p^q(\alpha) \delta e^p(\alpha) \right)}{\partial x_q} dx.$$

According to the gravitational equations the bracket in the first integral $= -\kappa \mathfrak{t}_p(\alpha)$, the integral itself

$$= -\kappa \int \mathfrak{t}_q(\alpha) \frac{\partial e^q(\alpha)}{\partial x_q} \xi^p dx.$$

One introduces

$$\mathfrak{v}_p^q = \delta_p^q \mathfrak{g} - \frac{\partial e^r(\alpha)}{\partial x_p} \mathfrak{g}_r^q(\alpha).$$

Equation (25) shows that the integral of

$$\left(\mathfrak{v}_p^q - \kappa \mathfrak{t}_q(\alpha) \frac{\partial e^q(\alpha)}{\partial x_p} \right) \xi^p$$

taken over an arbitrary volume of space-time is zero. The integrand must therefore vanish everywhere. Since the ξ^p are arbitrary constants the coefficients of the ξ^p are separately zero. In this way (21) is converted into a pure divergence equation

$$\frac{\partial \left(\mathfrak{v}_p^q + \kappa \mathfrak{t}_p^q \right)}{\partial x_q} = 0$$

and $\mathfrak{v}_q^p / \kappa$ is seen to be the gravitational energy. To formulate in addition a true differential conservation law for the angular momentum in general relativity one must choose coordinates such that the contragredient rotation of all tetrads appears as an orthogonal transformation of the coordinates. This is certainly possible, but I shall not pursue this question further here.

6. *Electric Field.* We come now to the critical part of the theory. In my opinion the origin and necessity for the electromagnetic field is the following. The components ψ_1, ψ_2 are, in fact, not uniquely determined by the tetrad but only to the extent that they can still be multiplied by an arbitrary "gauge-factor" $e^{i\lambda}$. The transformation of the ψ induced by a rotation of the tetrad is determined only up to such a factor. In special relativity one must regard this gauge-factor as a constant because here we have only a single point-independent tetrad. Not so in general relativity; every point has its own tetrad and hence its own arbitrary gauge-factor; because by the removal of the rigid connection between tetrads at

different points the gauge-factor necessarily becomes an arbitrary function of position. But then the infinitesimal linear transformation dE of the ψ, which corresponds to the infinitesimal rotation $d\Omega$, is not fully determined, but can be extended by an arbitrary pure imaginary multiple $i.df$ of the unit matrix. For the unique determination of the covariant differential $\delta\psi$ of ψ one needs, in addition to the metric in the neighbourhood of the point P, such a df for every line element $PP' = (dx)$ radiating from P. In order that $\delta\psi$ still depend linearly on dx

$$df = f_p(dx)^p$$

must be a linear form in the components of the line-element. If one replaces ψ by $e^{i\lambda}\psi$ one must, according to the covariant differential formalism, replace df by $df - d\lambda$ at the same time.

The result of this is that the term

$$\frac{1}{\epsilon}f(\alpha)s(\alpha) = \frac{1}{\epsilon}f(\alpha).\psi^* S(\alpha)\psi = f_p.\psi^*\mathfrak{S}^p\psi \tag{26}$$

must be added to the action-density $\mathfrak{m}$. From now on $\mathfrak{m}$ will denote the complete action density. We then have gauge-invariance in the sense that the action remains unchanged under the replacements

$$\psi \quad \text{by} \quad e^{i\lambda}\psi, \qquad f_p \quad \text{by} \quad f_p - \frac{\partial\lambda}{\partial x_p},$$

where λ is understood to be an arbitrary function of position. It is exactly in the manner described by (26) that the electromagnetic potential interacts with matter according to experiment. This justifies the identification of the quantities f_p introduced here with the electromagnetic potentials. The proof is complete if we show that, conversely, the f_p-field is influenced by the matter according to the laws that are shown by observation to be valid for the electromagnetic field.

$$f_{pq} = \frac{\partial f_q}{\partial x_p} - \frac{\partial f_p}{\partial x_q}$$

is a gauge-invariant anti-symmetric tensor and

$$\mathfrak{l} = \frac{1}{4}f_{pq}\mathfrak{f}^{pq} \tag{27}$$

the scalar density characteristic of the Maxwell field. The Ansatz

$$\mathfrak{h} = \mathfrak{m} + a\mathfrak{l} \tag{28}$$

(a a numerical constant) yields by variation of f_p the Maxwell equations with

$$-\,\mathfrak{s}^p = -\psi^*\mathfrak{S}^p\psi \tag{29}$$

as the electrical 4-current density.

The gauge-invariance is closely connected with the conservation of electric charge. Since $\mathfrak{h}$ is gauge-invariant $\delta\int\mathfrak{h}dx$ must vanish identically when for fixed $e^p(\alpha)$ the ψ and f_p are varied according to

$$\delta\psi = i\lambda\psi, \qquad \delta f_p = -\frac{\partial\lambda}{\partial x_p},$$

λ being an arbitrary function of position. This produces a relation between the material and electromagnetic equations which is identically satisfied. If we know that the field-equations for the matter (in the narrow sense) are valid it follows that

$$\delta\int\mathfrak{h}.dx = 0,$$

when the f_p alone are varied according to $\delta f_p = -\partial\lambda/\partial x_p$. On the other hand the same result follows from the equations of electromagnetism for the infinitesimal variation $\delta\psi = i\lambda.\psi$ of the ψ alone. If $\mathfrak{h} = \mathfrak{m} + a\mathfrak{l}$ one obtains in each case

$$\int\delta\mathfrak{h}.dx = \pm\int\psi^*\mathfrak{S}^p\psi\frac{\partial\lambda}{\partial x_p}.dx = \mp\int\lambda\frac{\partial\mathfrak{s}^p}{\partial x_p}dx.$$

We found earlier an analogous situation for the conservation of energy-momentum and angular momentum. They connected the field-equations for matter in the wider sense with the gravitational equations and corresponded to invariance with respect to coordinate transformations or equivalently with respect to independent rotations of the local tetrads at different points. From

$$\frac{\partial\mathfrak{s}^p}{\partial x_p} = 0 \tag{30}$$

it follows that the flow of the vector-density s^p through a three-dimensional section of space-time, in particular through a section (22)

$$1 = \int\mathfrak{s}^o d\xi \tag{31}$$

is independent of the position of the section, respectively t. Not only the integral, but each separate integral-element has an invariant meaning; however the sign depends on which direction is taken as the positive direction of the three-dimensional section. For $\mathfrak{s}^0 d\xi$ to be a space-like probability-density the hermitian form

$$e^o(\alpha)\psi^* S(\alpha)\psi \tag{32}$$

of ψ_1, ψ_2 must be definite. One easily finds that this is the case if (22) is actually a space-like section through P i.e. if the line-elements originating in P are space-like. In order to obtain the positive sign in (32) the sections $x_0 =$ constant, ordered according to increasing x_0, should follow the future direction determined by the vector $\mathbf{e}(0)/i$. With this natural restriction of the coordinate-system the sign of the current is determined and the invariant (31) is normalized in the usual way by the condition

$$\mathfrak{I} \equiv \int \mathfrak{s}^o d\xi = 1 \tag{33}$$

The constant a that connects $\mathfrak{m}$ with $\mathcal{L}$ is then a pure number $= ch/e^2$ (reciprocal of the fine-structure constant).

We treat $\psi_1, \psi_2;\ f_p;\ e^p(\alpha)$ as the mutually independent quantities to be varied. On account of the additional term in (26) the energy density $\mathfrak{t}_q^p$ coming from $\mathfrak{m}$ must be extended by

$$f_p \mathfrak{s}^q - \delta_p^q (f_r \mathfrak{s}^r).$$

In special relativity this leads to the energy being associated with the operator

$$H = \sum_{p=1}^{3} S^p \left(\frac{1}{i} \frac{\partial}{\partial x_p} + f_p \right)$$

since it has the value

$$\int \psi^* H \psi d\xi.$$

Of course, the matter equations then take the form

$$\left(\frac{1}{i} \frac{\partial}{\partial x_o} + f_o \right) \psi + H\psi = 0 \quad \text{and not} \quad \frac{1}{i} \frac{\partial \psi}{\partial x_o} + H\psi = 0,$$

as was formerly assumed in quantum-mechanics. Naturally, the matter-energy must be supplemented by the electromagnetic energy, for which Maxwell's classical expression is still valid.

As far as physical dimensions are concerned, it is natural in general relativity to take the x_p to be pure numbers. The quantities that appear are not only invariant with respect to scale-changes but with respect to arbitrary variations of the x_p. If we make the change $\mathbf{e}(\alpha) \rightarrow b\mathbf{e}(\alpha)$ by multiplication with a constant b then to preserve the normalization the ψ must be changed to $b^{3/2}\psi$. Thereby $\mathfrak{m}$ and $\mathfrak{l}$ are not changed and are therefore pure numbers. In contrast, $\mathfrak{g}$ acquires the factor $1/b^2$ so that κ is the square of a length d. κ is not identical with Einstein's gravitational constant but is obtained from it by multiplication with $2h/c$. The constant d is much smaller than the atomic order of magnitude, it is $\sim 10^{-32}$. Thus here also gravitation will be significant only for astronomical problems.

Thus, if we ignore gravity, the field-equations contain no dimensional atomic constant. In the two-component theory there is no place for a term in the action like the mass-term in the Dirac theory [6]. But one knows how the mass can be introduced on the basis of the conservation laws. One assumes that in the "empty environment" of the particle, outside a certain space-time tube whose x_0 = constant section is finite, the $\mathfrak{t}_p^q$ vanish and the $e^p(\alpha)$ take the constant values of special relativity. Then

$$J_p = \int \left(\mathfrak{t}_p^o + \frac{1}{\kappa}\mathfrak{v}_p^o \right) d\xi$$

are the components of a four-vector which is constant in time and is not influenced by the choice of coordinate system or the local tetrad. The normal coordinate system can be fixed more precisely by the condition that the three-momentum (J_1, J_2, J_3) vanish; then $-J_0$ is the invariant and simultaneously the constant mass of the particle. It is then required that the value of this mass be given once and for all.

In addition to the above-mentioned theory of the electromagnetic field, which I regard as correct because it originates so naturally in the arbitrariness of the gauge-factor in ψ and makes the connection between the experimentally observed gauge-invariance and charge-conservation understandable, there is another possible theory that connects electricity with gravitation. The term (26) has the same form as the second part of in (17); $\phi(\alpha)$ plays in the latter the same role as $f(\alpha)$ in the former. Hence one might expect that matter and gravitation, ψ and $e^p(\alpha)$, are already sufficient to explain electromagnetic phenomena, by considering the $\phi(\alpha)$ as the electromagnetic potentials. The dependence of these quantities on the $e^p(\alpha)$ and their first derivatives is such that they are invariant with respect to arbitrary coordinate transformations. But with respect to rotations of the tetrads the $\phi(\alpha)$ transform like fixed vectors of the tetrad only if the tetrads are subjected to the same rotations at all points. If one ignores the matter and considers only the relationship between electromagnetism and gravitation, one arrives at a theory of electromagnetism which is exactly of the same kind as that recently attempted by Einstein. However, in the present case the distant-parallelism is only simulated.

I have persuaded myself that the latter Ansatz, which at first sight is perhaps attractive, cannot yield the Maxwell equations. Furthermore the gauge-invariance would remain quite puzzling; the electromagnetic potentials and not merely the field-strengths would have a physical meaning. Thus I believe that this idea is a red-herring and that we should place much more trust in the hint that is provided by gauge-invariance: that electromagnetism is an accompanying phenomenon of the material wave-field and not of gravitation.

Palmer Physical Laboratory, Princeton University, 19, April 1929

[1] E. Wigner, ZS. f. Phys. **53**, 592, 1229; among others.

[2] In formal agreement with Einstein's recent work on Gravitation and Electricity, Sitzungsber. Preuss. Ak. Wissensch. 1928, p. 217, 224; 1929, p.2. Einstein uses the letter h instead of e.

[3] See for example H. Weyl, Space-Time-Matter 5th Ed., p. 233ff. (cited as RZM). Berlin 1923

[4] See RZM, p. 119f.

[5] See RZM, p. 272f.

[6] Proc. Roy. Soc. (A) **117**, 610.

PART II

The Nuclear Interactions

6

Klein's Serendipity, 1938

The attempts by Weyl and Kaluza to unify electromagnetism and gravitation ignored the existence of the other two fundamental forces, namely, the weak and strong nuclear forces, and what appears to be the first attempt to take these interactions into account was in a remarkable paper by Klein in 1938. This paper was presented at a Conference [1] on New Theories in Physics held at Kazimierz (Poland) in 1938 and was published in the proceedings of that conference.

The paper was presented two years after Yukawa's suggestion that the strong nuclear forces might be mediated by massive mesons in the same way that the atomic forces are mediated by photons. The suggested mesons would have to be massive to account for the short range of the nuclear forces, and some of them would have to be charged to account for the proton-neutron interactions. However, their tensorial character was not known, though they were generally assumed to be scalars or vectors. Proceeding on the assumption that they were vectors like the photon, Klein attempted to construct a unified theory of gravitation, electromagnetism and strong nuclear interactions. The paper was remarkable not only because it was the first to try to incorporate the Yukawa field into the usual gravitational and electromagnetic theories, but because it thereby produced a mathematical structure that would now be recognized as an $SU(2)$ gauge structure. Thus the paper was much ahead of its time, and, judging by the lack of discussion following Klein's presentation, the ideas were not fully appreciated by many of the eminent physicists who were present. Partly for this reason, partly because it was never published in a regular journal, and partly because of the outbreak of the Second World War soon afterwards, the paper went largely unnoticed.

The basic idea of the paper was to use the Kaluza theory in a form which was modified to take the meson-nucleon forces and the electric charges of the nucleons and mesons into account. The essential modifications introduced by Klein were:

(a) The fields were not assumed to be independent of the fifth coordinate x^o but to depend on it through a factor e^{-iex^o} where e is the electric charge. Thus $i\partial_o$ was effectively the electric charge operator.

(b) The 5-dimensional metric tensor was assumed to be of the form

$$g_{\mu\nu}(x) \qquad g_{oo} = 1 \qquad g_{\mu o}(x) = \beta\chi_\mu(x), \tag{136}$$

where $g_{\mu\nu}$ was the ordinary 4-dimensional metric field, β was a constant (which we normalize to unity), and $\chi_\mu(x)$ was a *matrix-valued* field of the form

$$\chi_\mu(x) = \begin{pmatrix} A_\mu(x) & \tilde{B}_\mu(x) \\ B_\mu(x) & A_\mu(x) \end{pmatrix} = \sigma_3(\vec{A}_\mu(x)\cdot\vec{\sigma}), \tag{137}$$

where $\vec{A}_\mu(x)\cdot\vec{\sigma}$ is the conventional $SU(2)$ gauge-potential. For c-number coordinates x^μ and x^o, eq. (137) implies that part of the the line-element is matrix-valued. The field A was identified with the electromagnetic potential and the fields $\tilde{B}$ and B with positively and negatively charged mesons, respectively.

It will be noticed that there is only one coupling constant, and thus the strength of nuclear interactions was assumed to be of the same order of magnitude as the electromagnetic interaction. Furthermore, since the mesons B and $\tilde{B}$ were charged the charge-independence of the strong nuclear forces was not taken into account (but more about this later). For later use we note that, from (a) above and the charge assignments,

$$i\partial_o\chi_\mu = e\begin{pmatrix} 0 & \tilde{B}_\mu \\ -B_\mu & 0 \end{pmatrix} = \frac{e}{2}[\sigma_3, \chi_\mu]. \tag{138}$$

As in Kaluza's theory, the coordinate transformations

$$x^o \quad \rightarrow \quad x^o + \alpha(x^\mu), \tag{139}$$

in the 5-dimensional space can be thought of as gauge transformations in the 4-dimensional space. In accordance with the usual rule for the variation of the metric tensor with respect to infinitesimal coordinate transformations (equation (67)) one finds that, for infinitesimal $\alpha(x)$, the variation of g_{oo} is zero, in agreement with the assumption that $g_{oo} = 1$, and that for $g_{oo} = 1$ the variation of the components $g_{\mu o}$ is

$$\delta g_{\mu o} = (\partial_\mu + g_{\mu o,o})\alpha = \nabla_\mu\alpha \quad \text{where} \quad \nabla_\mu = \partial_\mu + \frac{ie}{2}[\chi_\mu, \sigma_3]. \tag{140}$$

From the second equation in (140) we see that, insofar as we regard (140) as a gauge transformation for the non-gravitational fields $g_{\mu o}$, the covariant derivative associated with it is ∇_μ. The corresponding covariant derivative for isospinor fields found by Klein is

$$D_\mu = \partial_\mu + ie\chi_\mu\left(\frac{1-\sigma_3}{2}\right). \tag{141}$$

Hence we should expect to see ∇_μ and D_μ rather than ∂_μ appearing in any dimensionally-reduced equation. Note that D_μ closely resembles the covariant derivative of $SU(2)$ gauge theory.

Klein first applied the theory to the nucleons, which he assumed to be the usual isospin doublet $\psi(x) = \begin{pmatrix} \psi_p(x) \\ \psi_n(x) \end{pmatrix}$. First, from the definition of Dirac matrices in curved space,

$$\{\gamma^a(x), \gamma^b(x)\} = 2g^{ab}(x) \quad \rightarrow \quad \{\gamma^o(x), \gamma^\mu(x)\} = 2g^{o\mu}(x) = 2\chi^\mu(x), \tag{142}$$

he obtained, modulo some approximations,

$$\gamma^o(x) = \gamma_\nu \chi^\nu(x), \tag{143}$$

where γ_ν is the usual flat-space Dirac matrix. Using this result, he was able to write the 5-dimensional Dirac Lagrangian,

$$L^o = \bar{\psi}\left(\gamma^a(x)\partial_a + m\right)\psi = \bar{\psi}\left(\gamma^\nu(x)\partial_\nu + \gamma^o(x)\partial_o + m\right)\psi, \tag{144}$$

in the 4-dimensional form

$$L^o = \bar{\psi}\left(\gamma^\nu(x)D_\nu + m\right)\psi. \tag{145}$$

As expected, the derivative appears in the covariant form D_ν. On decomposing (145) into the terms corresponding to individual fields, one obtains Klein's equations (16) and (22), respectively. From Klein's equation (22), one can see how the charged mesons mediate a proton-neutron interaction.

Klein still had the problem of constructing a kinetic Lagrangian for the vector mesons. For this he followed the Kaluza procedure and assumed that it would be the 5-dimensional Einsteinian Action $\int \sqrt{g}R$ where R is the scalar curvature. For this action he used the same form as (59), namely,

$$\int \sqrt{g}R = \int \sqrt{g}g^{ab}\left\{{c \atop d[a}\right\}\left\{{d \atop c]b}\right\}, \tag{146}$$

in which the second-derivative terms in R have been removed by partial integration. On decomposing the expression (146) into its purely 4-dimensional and other parts, he obtained

$$\int \sqrt{g}\left[R_4 - \frac{1}{4}\mathrm{tr}\left(F_{\mu\nu}F^{\mu\nu}\right)\right] \quad \text{where} \quad F_{\mu\nu} = \nabla_\mu\chi_\nu - \nabla_\nu\chi_\mu, \tag{147}$$

and R_4 is the 4-dimensional scalar curvature.

This is the same expression as was obtained by Kaluza, except that the ordinary derivative ∂_μ has been replaced by the covariant derivative D_μ and

Kaluza's scalar field $g_{oo}(x)$ is missing because Klein had assumed it to be constant from the beginning. On computing the field strengths $F_{\mu\nu}$ explicitly, Klein obtained (see his equations (30) and (31))

$$F^A_{\mu\nu} = \partial_\mu A_\nu - \partial_\nu A_\mu + ie(B_\mu \tilde{B}_\nu - B_\nu \tilde{B}_\mu) \tag{148}$$

$$\begin{aligned} F^B_{\mu\nu} &= \partial_\mu B_\nu - \partial_\nu B_\mu + ie(A_\mu B_\nu - A_\nu B_\mu) \\ F^{\bar{B}}_{\mu\nu} &= \partial_\mu \bar{B}_\nu - \partial_\nu \bar{B}_\mu - ie(A_\mu \bar{B}_\nu - A_\nu \bar{B}_\mu). \end{aligned} \tag{149}$$

By writing the gauge potentials in the $SU(2)$ notation

$$\vec{A}_\mu \equiv \{A^1_\mu, \quad A^2_\mu, \quad A^3_\mu\} = \left\{ \frac{(B_\mu + \tilde{B}_\mu)}{\sqrt{2}}, \quad \frac{(B_\mu - \tilde{B}_\mu)}{i\sqrt{2}}, \quad A_\mu \right\}, \tag{150}$$

respectively, we see that (148) and (149) can be combined into

$$F^a_{\mu\nu} = \partial_\mu A^a_\nu - \partial_\nu A^a_\mu + e\epsilon_{abc} A^b_\mu A^c_\nu, \tag{151}$$

which are just the $SU(2)$ expressions for the field strengths.

One may ask how Klein, who was not even thinking of non-abelian gauge theory, used only one extra dimension, and gave special preference to the electromagnetic field, could arrive at an $SU(2)$ gauge Lagrangian.

The answer is the following: The expressions for $F^B_{\mu\nu}$ and $F^{\bar{B}}_{\mu\nu}$ in (149) are actually the electromagnetic covariant derivatives of the B fields and are therefore an automatic consequence of electromagnetic gauge invariance. So the question reduces to why the expression for F^A contains a bilinear in the **B** fields, in contrast to most other theories of charged vector fields, for which F^A is simply the curl of A. The reason is that the dimensional reduction replaces the ordinary curl by the covariant curl shown in (140), and the latter contains the commutator $[\chi_\mu, \partial_o \chi_\nu]$, which is not zero because χ_μ is matrix-valued. In fact, from (138) we have

$$\begin{aligned} F^A_{\mu\nu} &= \frac{1}{2} \operatorname{tr} \sigma_3 (\nabla_\mu \chi_\nu - \nabla_\nu \chi_\mu) = \partial_\mu A_\nu - \partial_\nu A_\mu + \frac{e}{2i} \operatorname{tr}([\sigma_3, \chi_\mu] \chi_\nu) \\ &= \partial_\mu A_\nu - \partial_\nu A_\mu + \frac{ie}{2} \operatorname{tr}(\sigma_3 [\vec{A}_\mu \cdot \vec{\sigma}, \vec{A}_\nu \cdot \vec{\sigma}]) \end{aligned} \tag{152}$$

which is just the first equation in (149).

Since the vector mesons B_μ have to be massive (with mass μ, say) in order to make the nuclear forces short-range, Klein added a mass term of the form $\mu^2 g^{\mu\nu} B_\mu \bar{B}_\nu$ to his Lagrangian. This did not violate any of his principles; in particular, it preserved the electromagnetic gauge invariance, but it did, of course, break what we would now call the $SU(2)$ gauge invariance. Since Klein was not demanding $SU(2)$ invariance, he was not concerned by this. But he was concerned by the ad hoc nature of the mass terms and remarked that " it is not impossible that a further development of the theory will make this somewhat arbitrary addition superfluous, the mass appearing as some sort of self-energy determined by other lengths entering in the theory," thus foreshadowing the Higgs mechanism.

Using the Lagrangians (145) and (146), Klein then proceeded to derive the usual field equations and conservation laws. He also wrote down the formal equal-time canonical quantization rules. Thus, already in 1938, Klein had stumbled onto the $SU(2)$ gauge theory (softly broken by the B masses).

After Klein's talk the only comment was by C. Moller, who pointed out that the charge-independence of the strong nuclear forces required that the **B** fields be accompanied by a neutral vector meson and asked how such a meson could be accommodated by Klein's theory. No problem, answered Klein, all one has to do is to change the matrix field χ_μ of (132) to

$$\chi_\mu = \begin{pmatrix} A_\mu - C_\mu & \tilde{B}_\mu \\ B_\mu & A_\mu + C_\mu \end{pmatrix} \quad \text{or} \quad \chi_\mu = \sigma_3 \left(\vec{A}_\mu . \vec{\sigma} - C_\mu \right) \tag{153}$$

Thus, although he was not thinking in such terms, Klein effectively generalized the Lie algebra of the vector meson fields from $SU(2)$ to $SU(2) \times U(1)$. He thereby anticipated the gauge group used in the standard model of electroweak theory by more than twenty years!

It should be noted, however, that if one were to pursue Klein's four-vector-field proposal, the identification of the A_μ and C_μ fields would have to be interchanged because the A_μ field couples (strongly) to the B_μ fields. Hence C_μ rather than A_μ should be identified as the electromagnetic field.

Finally, with respect to the weak interactions, the following comment of Klein is perhaps worth noting:

> [T]he Lagrangian L^o may belong either to the pair proton-neutron or to the pair neutrino-electron . . . the complete Lagrangian will imply an interaction of heavy and light spinor particles not only through the . . . electromagnetic field but also through the B-field, an interaction that will entail the occurrence of β processes.

Thus, by switching to the weak nuclear interactions, he also anticipated the general form of the present electroweak theory.

On the Theory of Charged Fields

by Oskar Klein in Copenhagen

Report submitted to Conference on New Theories in Physics, Kazimierz, Poland, 1939

I. —Introduction

The discovery of the so called heavy electron or mesoton and the role it is supposed to play in nuclear physics for the occurence of attractive forces at small distances—a role suggested by Yukawa [1] already before the discovery—would seem to mean a considerable enlargement of the region of applicability of the field concept, which has hitherto been limited by the self energy difficulties. In fact, while electrons could be treated unambiguously only down to distances large compared to the so called radius of the electron, the new particle would seem to present no principal difficulties before one approaches its own "radius", which, due to its larger mass, is about two orders of magnitude smaller than that of the electron. Moreover a characteristic length of just the order of magnitude of the electronic radius is introduced as a "Compton" wave length corresponding to the mass of the new particle, which, as shown by Yukawa, gives the range of the forces due to the corresponding field.

The logical consistency of this enlargement of the field concept would seem to require the removal of the self energy difficulty of the electron at least down to distances approaching the radius of the new particle. Considering the order of magnitude and range of nuclear forces—the nuclear binding energy being comparable with the rest mass energy of the electron and the range of the forces of the order of the electronic radius—it would not seem unreasonable to assume that a theory explaining nuclear attractions would also account for the rest mass of the electron, the attractive forces required as a compensation of the Coulomb repulsion being of a similar nature as the nuclear forces. A necessary condition for such an explanation—which would mean a removal of at least two orders of magnitude of the point from where the higher frequencies are "cut off"—is that the new forces belonging to the heavy electron field are determined by means of the elementary electric charge in a similar way as the electromagnetic forces, so that no other independent constant than the mass of the new particle will appear in the theory. Here we shall not enter further on the interesting problem as to the uniqueness of such a field theory comprising also the charged fields corresponding to the new particle if the demands of the electronic self energy and those of general invariance are to be satisfied, but we shall show how a theory of this kind may be built up as a consequent generalization of the formal scheme of ordinary field theory. If this theory of the mesoton and its interaction with electromagnetic fields and ordinary particles prove correct, it ought to satisfy automatically the above demands as to the self energy.

For the invariant formulation of the theory the so called five dimensional representation has proved useful as a very direct and simple means of expressing the fundamental conservation theorems for energy, momentum and electric charge and their relation to space-time translation invariance and gauge invariance. In the original form of that representation, which leads exactly to the Einstein-Maxwell theory of gravitation and electromagnetism, the field quantities are supposed not to contain the new auxiliary coordinate x^o. The fact that this coordinate appears as the canonical conjugate (apart from a constant factor) to the electric charge suggests, however, as a natural generalization the assumption that the field quantities contain also terms depending upon x^o and representing charged fields which resemble those represented by the solutions of the quantum theoretical wave equations of electric particles [2]. The existence of a smallest quantum of electricity would hereby require—as a classical model later to be replaced by a suitable quantization—a periodicity in x^o corresponding to the length $l_o = \frac{he\sqrt{2\kappa}}{c}$ where h is Planck's quantum of action, c the vacuum velocity of light, e the elementary quantum of electricity and κ the Einstein gravitational constant [3]. The period being invariant under a gauge transformation of x^o, this model will satisfy automatically the claim of gauge invariance.

The direct and general way it expresses the fundamental conservation and invariance theorems seems to make this representation a natural starting point for a general quantum field theory comprising also the charged fields, which are supposed to correspond to the mesotons. The field quantities, which may be taken in a form adapted to the demands of a generally relativistic Dirac wave equation in five dimensions [4], would moreover have to satisfy commutation relations corresponding to the Bose-Einstein statistics. While these quantities will represent particles with integer spin, the ordinary elementary particles would have to be represented by Dirac spinor wave functions. But just as the field quantities contain parts depending upon x^o representing charged fields and parts without x^o representing the ordinary electromagnetic and gravitational fields, we shall assume that also the spinors consist of x^o-free components representing neutral particles (neutron, neutrino) and of x^o components representing charged particles (proton, electron). Probably the higher harmonics of the Fourier development with respect to x^o, which would correspond to multiply charged particles, have no physical significance. They may be avoided in a similar way as in the theory of electronic spin through the introduction of two-row matrices. This description of charged and neutral particles would be consistent with the way protons and neutrons are treated in Heisenberg's theory of nuclear constitution, the matrix representation just mentioned corresponding to the so called istopic spin. On the other hand this matrix representation is nearly connected with the gauge transformation.

The theory outlined, which may be derived from a variation principle, the Lagrangian containing the spinor as well as the tensor field quantities, will describe an interaction between the proton, the neutron, the electron and the neutrino through the intermediate of the charged and uncharged fields giving as it would seem a quantitative formulation of the considerations of Yukawa, Kemmer and Bhabha on nuclear fields of force [5]. Especially the theory will contain no new physical constants other than the mass of the mesoton, and the nuclear interactions will, like the electromagnetic forces in the first approximation neglecting all direct gravitational effects, depend upon e^2.

As to the rest mass of the new particle, which does not appear in the ordinary field equations, it might be introduced by the addition of a term to the Lagrangian without disturbing the invariance. But it is not impossible that a further development of the theory will make this somewhat arbitrary addition superfluous, the mass appearing as some sort of self energy determined by the other lengths entering in the theory.

As a lack of simplicity of the theory it should be mentioned that although the allowed transformations are limited to general four-dimensional co-ordinate transformations and gauge transformations it makes use of invariants and tensors belonging to a general five-dimensional Riemann space. In the older form of the theory, where the physical quantities do not depend upon x^o, this inconvenience has been avoided by means of a projective treatment [6]. Even if it may seem doubtful whether this treatment could be extended to the generalized form of the theory developed below, this lack is probably more formal than real. As to more or less arbitrary addition (in the field equations for the charged fields) and omission (in the Dirac equations) of mass terms connected with electric charges it may have to do with a real limitation of the theory, these terms being perhaps—as indicated above—related to different stages of the self energy problem. Whether the circumstance that the unitary treatment of gravitational, electromagnetic and charged fields yielded by the theory does not also comprise the spinor particles, the Lagrangian of which is simply added to the common Lagrangian of the fields mentioned, must be regarded as a lack is perhaps doubtful. It should further be emphasized that although gravitational actions will be neglected in the following the structure of the theory is very essentially determined by the circumstance that it fulfils from the outset the claim of general relativity. This indirect effect of gravitation may perhaps be compared with the influence of spin on the constitution of atoms and molecules also in the approximation where all direct effects of the spin momenta are neglected.

II. —Mathematical Treatment of the Spinor Particles

Turning now to the mathematical formulation of the above remarks we introduce a Riemann metric tensor $\gamma_{\mu\nu}$, μ and ν taking the values 0, 1, 2, 3, 4. Corresponding to the invariance towards general transformations of the space-time co-ordinates $x^1, x^2, x^3, x^4 = ict$ together with that towards the gauge transformation expressed by means of the transformation equation $x'^o = x^o +$ arbitrary function of space-time co-ordinates (A), we define this tensor through the following relations

$$\gamma_{oo} = 1, \quad \gamma_{ok} = \beta\chi_k, \quad \gamma^{kl} = g^{kl}, \quad k, l = 1, 2, 3, 4 \tag{1}$$

Here the χ_k are certain functions of x^o, x^1, x^2, x^3, x^4, to be more closely characterized below, β a constant equal to $\sqrt{2\kappa}$ and g^{kl} the 10 contravariant components of the Einstein metric tensor. Further we introduce a set of generally relativistic Dirac matrices γ_μ satisfying the relations

$$\frac{1}{2}(\gamma_\mu\gamma_\nu + \gamma_\nu\gamma_\mu) = \gamma_{\mu\nu} \tag{2}$$

and the corresponding quantities $\gamma^\mu = \gamma^{\mu\nu}\gamma_\nu$, which fulfill the relations

$$\frac{1}{2}(\gamma^\mu\gamma_\nu + \gamma_\nu\gamma^\mu) = \delta_{\mu\nu} \tag{3}$$

and

$$\frac{1}{2}(\gamma^\mu\gamma^\nu + \gamma^\nu\gamma^\mu) = \gamma^{\mu\nu}$$

The γ_μ and the γ^μ may be taken as linear combinations of 5 ordinary constant Dirac matrices $\varepsilon_o, \varepsilon_1, \varepsilon_2, \varepsilon_3, \varepsilon_4$ satisfying the relations

$$\frac{1}{2}(\varepsilon_\mu\varepsilon_\nu + \varepsilon_\nu\varepsilon_\mu) = \delta_{\mu\nu} \tag{4}$$

and since $\gamma_{oo} = 1$, we may put

$$\gamma_o = \varepsilon_o \tag{5}$$

Further it follows from the definition of the γ^μ that

$$\gamma_o = \gamma_{o\nu}\gamma^\nu = \gamma^o + \beta\chi_k\gamma^k$$

or

$$\gamma^o = \varepsilon_o - \beta\chi_k\gamma^k \tag{6}$$

where the summation with respect to k goes from 1 to 4.

We shall now consider the following expression

$$\phi\,\gamma^{\rho}\psi_{i\rho} - \phi_{i\rho}\gamma^{\rho}\psi \tag{7}$$

well-known from the Lagrangian leading to the generally relativistic form of Dirac's wave equations, the summation with respect to ρ being here extended from 0 to 4, where

$$\psi_{i\rho} = \frac{\partial\psi}{\partial x^{\rho}} - \Gamma_{\rho}\psi, \quad \phi_{i\rho} = \frac{\partial\phi}{\partial x^{\rho}} + \phi\Gamma_{\rho} \tag{8}$$

are so called covariant derivatives of ψ and ϕ respectively, the Γ_{ρ} being certain matrices, the necessity of which was first shown by Schrödinger. Introducing the expression (6) for γ^{o} and omitting the term $\phi\varepsilon_{o}\frac{\partial\psi}{\partial x^{o}} - \frac{\partial\phi}{\partial x^{o}}\varepsilon_{o}\psi$—which omission does not violate the invariance required although from a formal point of view it is not altogether satisfactory—we get

$$\phi\,\gamma^{k}(\nabla_{k}\psi) - (\nabla_{k}\phi)\gamma^{k}\psi - \phi(\gamma^{\rho}\Gamma_{\rho} + \Gamma_{\rho}\gamma^{\rho})\psi \tag{7a}$$

where

$$\nabla_{k} = \frac{\partial}{\partial x^{k}} - \beta\chi_{k}\frac{\partial}{\partial x^{o}} \tag{9}$$

From the assumption made above (and to be stated more precisely below) about the x^{o} dependence of the quantities concerned, it follows that $\beta\frac{\partial}{\partial x^{o}}$ will not contain β* and will have to be retained when—as now we shall do—all terms in the expression (7a) containing β or any power of it are omitted, which means that we neglect all direct gravitational effects. Using a Cartesian co-ordinate system ($g_{kl} = \delta_{kl}$) only Γ_{o} will be "finite" with this approximation and given by

$$\Gamma_{o} = \frac{1}{2}\beta\frac{\partial\chi_{k}}{\partial x^{o}}\varepsilon_{o}\varepsilon_{k};$$

but the term $\gamma^{o}\Gamma_{o} + \Gamma_{o}\gamma^{o}$ will be small of order β since ε_{o} anticommutes with all four ε_{k}, and we are left with the expression

$$\phi\,\gamma^{k}(\Delta_{k}\psi) - (\Delta_{k}\phi)\gamma^{k}\psi \tag{7b}$$

only, which by a suitable choice of the γ^{k} will also be correct in a non-Cartesian co-ordinate system, even in the non-Euclidean case [7].

As a Lagrangian L^{o} for our spin particles we take now

$$L^{o} = -\hbar_{c}[\phi\,\gamma^{k}(\nabla_{k}\psi) - (\nabla_{k}\phi)\gamma^{k}\psi] + 2Mc^{2}\phi\,\psi \tag{10}$$

*Missing from original; it is presumed that β is intended.

where the mass M of the particle is to be taken very nearly equal to the proton mass for the pair proton-neutron and probably equal to 0 for the pair electron-neutrino.

In a Cartesian co-ordinate system we may put

$$\gamma^k = \varepsilon_k \tag{11}$$

and further

$$\phi = \psi^* \varepsilon_4 \tag{12}$$

ψ^* being the conjugate complex wave function to ψ.

The more general expression (10) may be used in a well-known way to derive the components T^o of the corresponding energy-momentum tensor, which—since L^o will have the numerical value 0—may be defined as by means of the relation

$$\delta L^o = -T^o_{kl}\delta g^{kl}$$

obtained by putting

$$\delta\gamma^k = \frac{1}{2}\gamma_l \delta g^{kl} \tag{13}$$

This gives

$$T^o_{kl} = \frac{\hbar c}{4}\left[\phi\,\gamma_l(\nabla_k\psi) + \phi\,\gamma_k(\nabla_l\psi) - (\nabla_k\phi)\gamma_l\psi - (\nabla_l\phi)\gamma_k\psi\right] \tag{14}$$

where in a Cartesian co-ordinate system

$$\gamma_k = \varepsilon_k \tag{11b}$$

In conformity with what has been said above about the x^o-dependence of the quantities concerned we put now

$$\chi_k = \begin{pmatrix} A_k, & \tilde{B}_k \\ B_k, & A_k \end{pmatrix}, \qquad \beta\frac{\partial \chi_k}{\partial x^o} = \frac{ie}{\hbar c}\begin{pmatrix} 0, & -\tilde{B}_k \\ B_k, & 0 \end{pmatrix} \tag{15}$$

$$\psi = \begin{pmatrix} \psi_N \\ \psi_P \end{pmatrix}, \qquad \beta\frac{\partial \psi}{\partial x^o} = \frac{ie}{\hbar c}\begin{pmatrix} 0 \\ \psi_P \end{pmatrix}$$

$$\phi = (\phi_N, \phi_P), \qquad \beta\frac{\partial \phi}{\partial x^o} = -\frac{ie}{\hbar c}(0, \phi_P), \tag{15a}$$

χ_1, χ_2, χ_3 being Hermitian matrices and χ_4 a Hermitian matrix multiplied by i. The A_k will be seen to be the components of the electromagnetic potential four-vector, while the B_k and $\tilde{B}_k$ describe the charged field. After a simple calculation we get from (10) and (13)

$$L^o = -\hbar c \Bigg[\phi_N \gamma^k \frac{\partial \psi_N}{\partial x^k} - \frac{\partial \phi_N}{\partial x^k} \gamma^k \psi_N + \phi_P \gamma^k \frac{\partial \psi_P}{\partial x^k} - \frac{\partial \phi_P}{\partial x^k} \gamma^k \psi_P - 2\frac{ie}{\hbar c} A_k \phi_P \gamma^k \psi_P - \frac{ie}{\hbar c}(\tilde{B}_k \phi_N \gamma^k \psi_P + B_k \phi_P \gamma^k \psi_N) + 2Mc^2(\phi_N \psi_N + \phi_P \psi_P) \Bigg] \quad (16)$$

$$T^o_{kl} = \frac{\hbar c}{4} \Bigg[\phi_N \gamma_k \frac{\partial \psi_N}{\partial x^l} + \phi_N \gamma^l \frac{\partial \psi_N}{\partial x^k} - \frac{\partial \phi_N}{\partial x^l} \gamma_k \psi_N - \frac{\partial \phi_N}{\partial x^k} \gamma_l \psi_N + \phi_P \gamma_k \frac{\partial \psi_P}{\partial x^l} + \phi_P \gamma_l \frac{\partial \psi_P}{\partial x^k} - \frac{\partial \phi_P}{\partial x^l} \gamma_k \psi_P - \frac{\partial \phi_P}{\partial x^k} \gamma_l \psi_P - 2\frac{ie}{\hbar c}(A_k \phi_P \gamma_l \psi_P + A_l \phi_P \gamma_k \psi_P) - \frac{ie}{\hbar c}\left(\tilde{B}_k \phi_N \gamma_l \psi_P + \tilde{B}_l \phi_N \gamma_k \psi_P + B_k \phi_P \gamma_l \psi_N + B_l \phi_P \gamma_k \psi_N\right) \Bigg] \quad (17)$$

The expression (16) for L^o may further be used to determine the current-density four-vector s^{ok} since

$$S^{ok} = \frac{1}{2} \frac{\partial L^o}{\partial A_k} \quad (18)$$

In this way we get

$$S^{ok} = ie\, \phi_P \gamma^k \psi_P \quad (19)$$

In a Cartesian co-ordinate system this gives according to (12)

$$S^{ok} = ie\, \psi_P^* \varepsilon_4 \varepsilon_k \psi_P \quad (20)$$

and especially

$$S^{o4} = ie\, \psi_P^* \psi_P$$

in conformity with Dirac's original assumption about $\psi^*\psi$. We see also that only ψ_P and not ψ_N contributes to the electric current and density.

From the expression (16) for L^o and by means of the variational principle

$$\delta \int L^o \sqrt{g}\, dx = 0, \quad g = |g_{kl}|, \quad dx = dx^1 dx^2 dx^3 dx^4 \quad (21)$$

the variations of ϕ and ψ vanishing at the border of the space-time region under consideration, we obtain the following wave equations for $\psi_N, \phi_N, \psi_P, \phi_P$, where for the sake of simplicity we have used a Cartesian co-ordinate system

$$\varepsilon_k \left(\frac{\partial \psi_N}{\partial x^k} - \frac{ie}{2\hbar c} \tilde{B}_k \psi_P \right) = \frac{Mc}{\hbar} \psi_N,$$

$$\left(\frac{\partial \phi_N}{\partial x^k} + \frac{ie}{2\hbar c} B_k \phi_P \right) \varepsilon_k = -\frac{M_c}{\hbar} \phi_N,$$

$$\varepsilon_k \left(\frac{\partial \psi_P}{\partial x^k} - \frac{ie}{\hbar c} A_k \psi_P - \frac{ie}{2\hbar c} B_k \psi_N \right) = \frac{Mc}{\hbar} \psi_P,$$

$$\left(\frac{\partial \phi_P}{\partial x^k} + \frac{ie}{\hbar c} A_k \phi_P + \frac{ie}{2\hbar c} \tilde{B}_k \phi_N \right) \varepsilon_k = -\frac{Mc}{\hbar} \phi_P \qquad (22)$$

which are seen to be of the Dirac type, the terms containing the B-field, which do not appear in the ordinary equations, describing the action on the spinor particles due to the charged fields or heavy electrons.

III. —Treatment of charged and neutral fields

In order to obtain field equations for the A- and B-fields we shall now regard the quantity Γ, which is the five-dimensional analogue to the Lagrangian of the Einstein gravitational field equations, namely

$$\Gamma = \gamma^{\mu\nu} \left[\begin{Bmatrix} \mu\tau \\ \sigma \end{Bmatrix} \begin{Bmatrix} \nu\sigma \\ \tau \end{Bmatrix} - \begin{Bmatrix} \mu\nu \\ \rho \end{Bmatrix} \begin{Bmatrix} \rho\tau \\ \tau \end{Bmatrix} \right] \qquad (23)$$

$\begin{Bmatrix} \mu\nu \\ \rho \end{Bmatrix}$ being the well-known bracket expressions

$$\begin{Bmatrix} \mu\nu \\ \rho \end{Bmatrix} = \frac{1}{2} \gamma^{\rho\tau} \left(\frac{\partial \gamma_{\tau\mu}}{\partial x^\nu} + \frac{\partial \gamma_{\tau\nu}}{\partial x^\mu} - \frac{\partial \gamma_{\mu\nu}}{\partial x^\tau} \right) \qquad (24)$$

To begin with we consider only the case where gravitation may be altogether neglected using a Cartesian co-ordinate system, where the Einstein metric tensor is given by

$$g_{kl} = \delta_{kl} \qquad (25)$$

and further we neglect all quantities containing higher powers of β than the second. Using the expressions (1) we get after a simple calculation

$$\Gamma = -\frac{\beta^2}{4} \chi_{rs} \chi_{rs} \qquad (26)$$

where

$$\chi_{rs} = \nabla_r \chi_s - \nabla_s \chi_r \tag{27}$$

∇_r being the operator defined in (9).

In a non-Cartesian system the corresponding formula will be (where we have still neglected those terms in g_{kl} which depend upon x^o)

$$\Gamma = -\frac{\beta^2}{4} g^{kr} g^{ls} \chi_{kl} \chi_{rs} + G \tag{28}$$

where G is the Einstein Lagrangian formed by means of the g_{kl} in a similar way as (23) is formed by means of the $\gamma_{\mu\nu}$. Apart from the terms containing x^o, where we have neglected all terms of gravitational order of magnitude, the formula (28) is exact and contains the well-known five-dimensional representation of the Maxwell-Einstein field theory, the χ_{rs} being apart from x^o-terms the components of the electromagnetic field six-vector. Although (28) has a generally relativistic form and therefore may be expected to give a correct description of the influence of ordinary gravitational fields it will probably not give a consequent treatment of the finer gravitational effects, since the neglected terms—among others these corresponding to charged g_{kl}-fields—are of the same gravitational order of magnitude. For our present purpose, however, which is to determine the energy-momentum tensor of the charged field in the non-gravitational case it will suffice.

Substituting now for the χ_r the expression (15) we get

$$\chi_{rs} = \begin{pmatrix} A_{rs}, & \tilde{B}_{rs} \\ B_{rs}, & A_{rs} \end{pmatrix} \tag{29}$$

where

$$A_{rs} = F_{rs} + \frac{ie}{\hbar c}(B_r \tilde{B}_s - B_s \tilde{B}_r) \text{ and } F_{rs} = \frac{\partial A_s}{\partial x^r} - \frac{\partial A_r}{\partial x^s} \tag{30}$$

F_{rs} being the components of the ordinary electromagnetic field six-vector and

$$\begin{aligned} B_{rs} &= \delta_r B_s - \delta_s B_r, \quad \delta_r = \frac{\partial}{\partial x^r} - \frac{ie}{\hbar c} A_r \\ \tilde{B}_{rs} &= \tilde{\delta}_r \tilde{B}_s - \tilde{\delta}_s \tilde{B}_r, \quad \tilde{\delta}_r = \frac{\partial}{\partial x^r} + \frac{ie}{\hbar c} A_r \end{aligned} \tag{31}$$

We have now to substitute the diagonal term of $\chi_{rs}\chi_{rs}$, namely $A_{rs}A_{rs} + B_{rs}\tilde{B}_{rs}$ into our Lagrangian (26) giving if we go over to the generally relativistic form

$$\Gamma = -\frac{\kappa}{2} g^{kr} g^{ls} (A_{kl} A_{rs} + B_{kl} \tilde{B}_{rs}) + G, \tag{32}$$

where we have replaced $\frac{\beta^2}{2}$ by κ. To obtain the total Lagrangian we have further to add a term corresponding to the rest mass μ of the mesoton, namely $-\kappa\frac{\mu^2c^2}{\hbar^2}g^{kl}\tilde{B}_kB_l$ and also the quantity κL^o which gives

$$L = \Gamma + \kappa L^o - \kappa\frac{\mu^2c^2}{\hbar^2}g^{kl}B_k\tilde{B}_l \tag{33}$$

In the variation principle

$$\delta\int L\sqrt{g}\,dx = 0 \tag{34}$$

the quantities to be varied are now $\psi_N, \phi_N, \psi_P, \phi_P, A_k, B_k, \tilde{B}_k$, and $g^{kl}, k, l = 1, 2, 3, 4$, all variations vanishing at the border of the region. The variation with respect to the $\psi : s$ and $\phi : s$ give, as we have already seen, the equations (22). The variation with respect to the g^{kl} gives us the Einstein field equations with right sides containing the energy-momentum tensor, namely

$$G_{kl} = R_{kl} - \frac{1}{2}g_{kl}R = -\kappa T_{kl} \tag{35}$$

where

$$T_{kl} = T^o_{kl} + T^e_{kl} \tag{36}$$

T^o_{kl} being the quantities defined by (17) while

$$T^e_{kl} = A_{kr}A_{lr} + \frac{1}{2}(B_{kr}\tilde{B}_{lr} + B_{lr}\tilde{B}_{kr}) - \frac{1}{4}\delta_{kl}(A_{rs}A_{rs} + B_{rs}\tilde{B}_{rs}) \tag{37}$$

R_{kl} are the components of the well-known Einstein curvature tensor and R the corresponding invariant. In the formula for T^e_{kl}, the components of the energy-momentum tensor due to the A- and B-fields, we have used for the sake of simplicity a Cartesian co-ordinate system. If the terms containing the $B : s$ are omitted, we obtain the usual formula for the electromagnetic energy-momentum tensor.

A variation of the A_k gives after a simple calculation, where again we use a Cartesian co-ordinate system,

$$\frac{\partial F_{kl}}{\partial x^l} = S_k = S^o_k + S^e_k \tag{38}$$

with

$$S^e_k = \frac{ie}{\hbar c}\left[\frac{\partial}{\partial x^l}(\tilde{B}_kB_l - \tilde{B}_lB_k) + \frac{1}{2}(\tilde{B}_{kl}B_l - B_{kl}\tilde{B}_l)\right] \tag{39}$$

while

$$S_k^o = ie\,\phi_P\,\varepsilon_k\,\psi_P \tag{19a}$$

is the spinor particle current given above. We see that equations (39) are the ordinary Maxwell equations for the electromagnetic field, the right side containing the current-density four-vector, to which the charged B-field contributes by the quantities S_k^e.

Finally we get wave equations for B_k and $\tilde{B}_k$ by varying B_k and $\tilde{B}_k$ respectively, namely

$$\begin{aligned} \delta_l\,B_{kl} - 2\frac{ie}{\hbar c}A_{kl}\,B_l + \frac{\mu^2c^2}{\hbar^2}B_k &= ie\,\phi_N\,\varepsilon_k\,\psi_P \\ \tilde{\delta}_l\,\tilde{B}_{kl} + 2\frac{ie}{\hbar c}A_{kl}\,\tilde{B}_l + \frac{\mu^2c^2}{\hbar^2}\tilde{B}_k &= ie\,\phi_P\,\varepsilon_k\,\psi_N \end{aligned} \tag{40}$$

We may easily verify the conservation theorem for the total electric charge. Indeed a simple calculation gives

$$\frac{\partial S_k^e}{\partial x^k} = \frac{ie}{2\hbar c}\left[B_l(\tilde{\delta}_k\,\tilde{B}_{kl}) - \tilde{B}_l(\delta_k\,B_{kl})\right]$$

or by means of (40)

$$\frac{\partial S_k^e}{\partial x^k} = \frac{e^2}{2\hbar c}(B_k\,\phi_P\,\varepsilon_k\,\psi_N - \tilde{B}_k\,\phi_N\,\varepsilon_k\,\psi_P) \tag{41}$$

From (22) we get similarly

$$\frac{\partial S_k^o}{\partial x^k} = -\frac{e^2}{2\hbar c}(B_k\,\phi_P\,\varepsilon_k\,\psi_N - \tilde{B}_k\,\phi_N\,\varepsilon_k\,\psi_P) \tag{41a}$$

showing that

$$\frac{\partial S_k}{\partial x^k} = \frac{\partial S_k^o}{\partial x^k} + \frac{\partial S_k^e}{\partial x^k} = 0$$

Thus the total charge is conserved, but not the charge of the spinor particles or that of the charged fields separately. At the same time the number of spinor particles (charged and uncharged) is conserved. In fact the sum of the probability current-density vectors belonging to the neutral and the charged spinor particles respectively, $i\,\phi_N\,\varepsilon_k\,\psi_N + i\,\phi_P\,\varepsilon_k\,\psi_P$ fulfils the continuity equation, in that

$$\frac{\partial}{\partial x^k}(i\,\phi_N\,\varepsilon_k\,\psi_N) = \frac{e}{2\hbar c}(B_k\,\phi_P\,\varepsilon_k\,\psi_N - \tilde{B}_k\,\phi_N\,\varepsilon_k\,\psi_P)$$

The equations (40) for the B-field are similar to though not identical with the equations given by Proca [8] and applied by several authors to the mesoton. The most important difference is the appearance of the non-linear terms in (40) due to the quantities $\frac{ie}{\hbar c}(\tilde{B}_r B_s - \tilde{B}_s B_r)$ in the A_{rs}. These same quantities are also seen to appear in the expression for the current-density vector, where they give rise to a magnetic spin moment. In fact, the component of the magnetic moment due to that part of the current in the direction of x^1 is given by the following expression

$$\frac{ie}{\hbar c}\int(\tilde{B}_2 B_3 - \tilde{B}_3 B_2)\,dr, \quad dr = dx^1 dx^2 dx^3,$$

where the integral has to be taken over all space, the expressions $\frac{ie}{\hbar c}(\tilde{B}_k B_l - \tilde{B}_l B_k)$ defining thus the density of magnetic spin moment [9].

It should perhaps be pointed out once more that the Lagrangian L^o may belong either to the pair neutron-proton or to the pair neutrino-electron. In (33) L^o should therefore, strictly speaking, be the sum of two different L^o : s, the one referring to the heavy and the other to the light spinor particles, each with its own set of ψ : s. This we have omitted for the sake of shortness and because the corresponding completion is quite obvious. But it is worthwhile to notice that the complete Lagrangian will imply an interaction of heavy and light spinor particles not only through the intermediate of the electromagnetic field but also through the B-field, an interaction which will entail the occurence of β-processes, the probability of which may be calculated on the basis of the theory developed in this report.

This theory, however, will not be completely founded before we have given rules for its quantization, a question we shall briefly touch upon here. First of all the ψ corresponding to the different kinds of spinor particles have to satisfy well-known quantization rules, being the application of the formalism of Jordan-Wigner to the wave equation of Dirac and corresponding to Fermi statistics. Moreover the electromagnetic quantities have to be quantized in the usual way, so that the remaining problem is the formulation of quantization rules for the B-field. Here we meet with the same formal difficulty as in the case of the electromagnetic field that the time derivatives of B_4 and $\tilde{B}_4$ are not present in the Lagrangian. We may, however, get over this difficulty by means of the general method developed by Rosenfeld. As a result one obtains the following commutation relations [10]

$$\begin{aligned}\left[\tilde{B}_{4k}(r),\ B_l(r')\right] &= 2\hbar c\,\delta_{kl}\,\delta(r-r') \\ \left[B_{4k}(r),\ \tilde{B}_l(r')\right] &= 2\hbar c\,\delta_{kl}\,\delta(r-r')\end{aligned} \tag{42}$$

where r and r' indicate two space points to be taken at the same time t, $\delta(r-r')$ being the well-known singular function. Further the B_k commute with all the

B_k, $\tilde{B}_k$ and all the B_{kl}, the $\tilde{B}_k$ similarly with all the B_k, $\tilde{B}_k$ and all the $\tilde{B}_{kl}$, while B_4 and $\tilde{B}_4$ commute with all the quantities mentioned.

As an interesting application of the commutation rules we may inquire into the commutation expression for the total charge Q^e of the B-field with one of the quantities B_k, $\tilde{B}_k$. We have

$$Q^e = -i \int S_4^e(r')\, dr' \tag{43}$$

S_4^e being given by (39). A simple calculation gives now

$$\left[Q^e,\, B_k(r)\right] = \frac{e}{2\hbar c} \int \left[\tilde{B}_{4k}(r'),\, B_k(r)\right] B_k(r')\, dr' = eB_k(r) \tag{44}$$

and similarly

$$\left[Q^e,\, \tilde{B}_k(r)\right] = -\frac{e}{2\hbar c} \int \left[B_{4k}(r'),\, \tilde{B}_k(r)\right] \tilde{B}_k(r)\, dr' = -e\tilde{B}_k(r) \tag{44a}$$

Since

$$\frac{\partial B_k}{\partial x^o} = \frac{ie}{\beta\hbar c} B_k, \qquad \frac{\partial \tilde{B}_k}{\partial x^o} = -\frac{ie}{\beta\hbar c} \tilde{B}_k,$$

we see that $\frac{Q^e}{\beta c}$ as far as the B-field is concerned plays the role of the canonical conjugate to x^o, which is a special case of a general rule according to which x^o is canonically conjugated to the total charge divided by βc. At the same time the relations (44) are seen to be closely related to the fact that the charge is always a whole positive or negative multiple of the elementary electric charge. They express moreover the gauge transformation properties of the B_k and $\tilde{B}_k$ [11].

Professor Moeller recalled that discussions of the experimental results relative to the interactions between the particles constituting nuclei appeared to show the existence of forces between two protons and between two neutrons of the same order of magnitude as the force between one proton and one neutron. If we wished to explain these forces by the intermediary of a field of particles capable of being given out or absorbed by protons and neutrons, we must then admit, in addition to Yukawa's charged particles, neuter particles of the same mass. It was hard to see how such neuter particles could naturally find a place in a general scheme of the type proposed by Professor Klein.

Professor Klein answered that if, in the Lagrangian, we replace the χ_{rs} given by (29), by the following expression

$$\chi_{rs} = \begin{pmatrix} A_{rs} - C_{rs}, & \tilde{B}_{rs} \\ B_{rs}, & A_{rs} + C_{rs} \end{pmatrix} \tag{29'}$$

where

$$C_{rs} = \frac{\partial C_s}{\partial x^r} - \frac{\partial C_r}{\partial x^s}$$

and, in the Lagrangian (10), the χ_r given by (15), by the corresponding expression,

$$\chi_r = \begin{pmatrix} A_r - C_r, & \tilde{B}_r \\ B_r, & A_r + C_r \end{pmatrix} \tag{15'}$$

The C-field corresponds to neuter particles to which it is possible to give any mass, for instance that of the mesoton. It seems, however, that such particles would give rise, according to the Lagrangian concerned, to repressive forces and not to attractive forces between protons.

[1] Yukawa 1935, Proc. Phys. Math. Soc. Japan **17**, 48; Yukawa et Sakata 1937, Proc. Phys. Math. Soc. Japan **19**, 1084; Yukawa, Sakata et Taketani 1938, Proc. Phys. Math. Soc. Japan **20**, 319, Yukawa, Sakata, Kobayasi, and Taketani, 1938, Proc. Phys. Math. Soc. Japan **20**, 720.

[2] Klein, Nature, 1926, **118**, 516, Z.f. Phys., 1927, **46**, 188.

[3] Klein, Nature, l.c.

[4] See Schrödinger, 1932, Berl. Ber., p.105 and Pauli and Solomon, 1932, Journal de Phys. (7), 3, 452 and 582.

[5] Yukawa, l.c.; Kemmer. Proc. Roy. Soc. A., **166**, 127, 1938; Fröhlich, Heitler et Kemmer, Proc. Roy. Soc. A., **166**, 154, 1937; Bhabha, Proc. Roy. Soc. A., **166**, 501, 1938.

[6] Voir W. Pauli, Ann. de Phys., 1933, **18**, 305 et 337.

[7] By a suitable choice of the γ_ρ we might have made $\gamma^\rho \Gamma_\rho + \Gamma^\rho \gamma_\rho$ equal to 0 from the beginning, but such a choice would in general not correspond to $\gamma_o = \varepsilon_o$.

[8] Proca, J. Phys. Radium, **7**, 347, 1936.

[9] See Proca, J. Phys. Radium, l.c.

[10] Similar relations for the Proca equations are given by Kemmer and Bhabha, Proc. Roy. Soc., l.c.

[11] See Heisenberg and Pauli, Z.f. Phys., 1930, **59**, 168.

7

Pauli's Dimensional Reduction, 1953

By the beginning of the fifties, the divergences that had plagued quantum field theory in Klein's time had been removed by renormalization, and the predictions of the renormalized quantum electrodynamics had been spectacularly verified by experiment. These developments, together with the better knowledge of the nuclear forces now available and the appearance of many new particles, placed the problem of finding a field theory for the strong nuclear interactions at the top of the agenda. At a conference [2] held in at Leiden in 1953, Pais proposed a field theory which took the charge-independence (isotopic-spin invariance) of the nuclear forces into account. This theory extended Weyl's ideas from charge conservation to isotopic spin conservation and prompted Pauli to make the following statement:

> I am very much in favour of the general principle to bring empirical conservation laws and invariance properties in connection with mathematical groups of transformations of Nature. If besides the conservation of energy-momentum and of charge the conservation of the property defined as the number of nucleons and charge-independence of the nuclear forces are well-established they have indeed, as Pais tried now to express mathematically, also to be connected with group theoretical properties of the laws of nature . . . I would like to ask in this connection whether the transformation group (isospin group) with constant phases can be amplified in a way analogous to the gauge-group for electromagnetic potentials in such a way that the meson-nucleon interaction is connected with the amplified group. The main problem seems to be the incorporation of the coupling constant into the group.

It is clear from these remarks how far Pauli had come from his first reaction to Weyl's 1929 paper. Later that year he wrote down the details of the sort of theory he had in mind "in order to see how it looks," as he said himself, and summarized his findings under the heading "Meson-Nucleon Interaction and Differential Geometry" in two letters that he wrote to Pais in July and December 1953.

The idea was that since the original Kaluza reduction produced the mathematical structure for describing electromagnetism, a generalization of that theory to more dimensions might produce the mathematical structure for describing the strong interactions. The letters contain the first correct expression for the non-abelian field strengths. The meson-nucleon interactions were introduced by means of the Dirac equation, and the most natural ones were found to be vectorlike, in contrast to the kind of interactions that were being proposed at the time (and which Pauli apparently disliked). Pauli did not write down a Lagrangian for the gauge fields, but it is clear from later remarks (see the next chapter) that he realized there would be a problem with the gauge-field masses.

Pauli's theory, like that of Kaluza and Klein, was based on dimensional reduction, but it differed from the Kaluza-Klein theory in two major respects. First, Pauli increased the number of extra dimensions from 1 to 2, thereby introducing non-abelian groups for non-gravitational interactions for the first time. Second, he identified the electromagnetic potentials with components of the Christoffel connection, whereas Kaluza and Klein had identied them with components of the metric tensor. The differences can be displayed symbolically by writing

$$\begin{array}{ccc} A_\mu = g_{o\mu} & & A^a_{\mu b} = \Gamma^a_{\mu b} \\ F_{[\mu\nu]} = \Gamma_{o[\mu\nu]} & \text{and} & F^a_{[\mu\nu]b} = R^a_{[\mu\nu]b}. \\ \text{(Kaluza)} & & \text{(Pauli)} \end{array} \tag{154}$$

It must be admitted that the Pauli proposal is more natural since it identifies two connections, rather than a connection and a tensor, and it has become the more standard one.

The Pauli letters are a little difficult to read, partly because of their exploratory character and partly because the approach adopted was not quite the same in both letters, which led to some overlap and some divergence. For this reason, and because the type of dimensional reduction used by Pauli was the prototype for most later reductions, we first summarize the general ideas involved and then return to the letters.

DIMENSIONAL REDUCTION

Consider a $(4+n)$-dimensional space with coordinates $x_A = (x_\mu, y_a)$ where the y's are the internal, or fiber-space, coordinates, and $y = 0$ is space-time . Let us restrict the possible coordinate transformations to coordinate transformations of space-time and linear transformations of the fiber-space which may depend on the space-time

$$y^a \quad \rightarrow \quad R^a_b(x) y^b \qquad a, b = 1 \dots n. \tag{155}$$

We shall refer to the transformations (155) as coordinate-gauge transformations since they are coordinate transformations in the full space but can be thought of as gauge transformations in space-time.

Let us now consider the parallel transfer of a vector $v(x)$ which depends only on the space-time coordinates, along a curve in space-time. Letting ∇ and $\{\}$ denote the usual Riemannian covariant derivative and Christoffel connection for the full space respectively, we find

$$(\nabla v_\mu)^\alpha = \partial_\mu v^\alpha + \left\{{\alpha \atop \mu\beta}\right\}_0 v^\beta + \left\{{\alpha \atop \mu b}\right\}_0 v^b$$
$$(\nabla v_\mu)^a = \partial_\mu v^a + \left\{{a \atop \mu b}\right\}_0 v^b + \left\{{a \atop \mu\beta}\right\}_0 v^\beta, \tag{156}$$

where the subscript zero means evaluation at $y = 0$. Suppose now that

$$\left(\partial_\mu g^a_\nu\right)_0 = \left(\partial_a g_{\mu\nu}\right)_0 = 0 \quad \text{or equivalently} \quad \left\{{\alpha \atop \mu b}\right\}_0 = \left\{{a \atop \mu\nu}\right\}_0 = 0, \tag{157}$$

the equivalence following immediately from the fact that the metric is covariantly constant. Then the cross-terms in (156) drop out and we obtain

$$(\nabla v_\mu)^\alpha = \partial_\mu v^\alpha + \left\{{\alpha \atop \mu\beta}\right\}_0 v^\beta \quad \text{and} \quad (\nabla v_\mu)^a = \partial_\mu v^a + \left\{{a \atop \mu b}\right\}_0 v^b, \tag{158}$$

which means that the the space-time components v^μ and the internal-space components v^a transform independently. Thus, under the assumption (157), the restriction of the parallel transfer of the full space to space-time decomposes into the ordinary parallel transfer in space-time for the space-time vectors v^ν and a parallel transfer of the form

$$D_\mu v^a = \partial_\mu v^a + A^a_{\mu b} v^b \quad \text{where} \quad A^a_{\mu b} = \left\{{a \atop \mu b}\right\}_0, \tag{159}$$

for the Kaluza-space vectors $v^a(x)$. The connection A_μ transforms as a vector with respect to coordinate transformations in space-time, but with respect to the coordinate-gauge transformations (155), it transforms according to

$$A^a_{\mu b} \to R^a_c R^d_b A^c_{\mu d} + R^c_a \partial_\mu R^c_b \quad \text{or} \quad A_\mu \to R^{-1} A_\mu R + R^{-1}\partial_\mu R, \tag{160}$$

where the inverse comes from the fact that R is orthogonal. This follows from the general transformation law for the metric components,

$$g_{ab} \to R^c_a R^d_b g_{cd} \quad \text{and} \quad g^a_\mu \to R^a_b g^b_\mu + \left(\frac{\partial R^a_c}{\partial x^\mu}\right) y^c, \tag{161}$$

and the definition

$$A^a_{\mu b} = \frac{1}{2} g^{ac} \left(\frac{\partial g_{\mu b}}{\partial y_c} - \frac{\partial g_{\mu c}}{\partial y_b} - \frac{\partial g_{bc}}{\partial x_\mu}\right)_0. \tag{162}$$

Equation (160) shows that $A^{a}_{\mu b}$, which is part of the Christoffel connection in the full Riemannian space, is a gauge connection from the point of view of space-time.

From (161) we see that the variation in the vector v^a on parallel transfer around an infinitesimal closed curve in space-time is

$$\delta v^a = R^a_{b\mu\nu} dx^\mu dx^\nu v^b, \tag{163}$$

where

$$\begin{aligned} R^a_{b\mu\nu} &= \left(\partial_\mu A_\nu - \partial_\nu A_\mu\right)^a_b + A^a_{\mu c} A^c_{\nu b} - A^a_{\nu c} A^c_{\mu b} \\ &= \left(\partial_\mu A_\nu - \partial_\nu A_\mu + [A_\mu, A_\nu]\right)^a_b . \end{aligned} \tag{164}$$

Hence $R^a_{b\mu\nu}$, which is part of the Riemann tensor for the full space, is also the Riemann tensor for the gauge connection, $A_\mu(x)$. This exhibits the relationship between the Riemann tensor of metrical geometry and the field strengths of gauge theory. From (163) one sees that that $R^a_{b\mu\nu} = 0$ is the necessary and sufficient condition that A_μ can be transformed to zero (locally) by a coordinate-gauge transformation (155).

It will be noticed that the formalism described here is very similar to the tetrad formalism. Indeed, it becomes identical if we choose the internal space to be four-dimensional and connect it to space-time by means of a tetrad.

THE PAULI LETTERS

Let us now discuss the original letters. Pauli's first step was to point out that the linear coordinate-gauge transformation (155) was the natural generalization of the Kaluza one, and the latter could be recovered from it by choosing $n = 2$ and regarding the Kaluza variable x^o as the angle on the unit circle. In other words,

$$x^o \to x^o + f(x_\mu) \quad \Leftrightarrow \quad y^a \to R^a_b(x) y^b \quad \text{where} \quad \begin{aligned} y^1 &= \sin(x^o) \\ y^2 &= \cos(x^o) \end{aligned}, \tag{165}$$

and $R \in SO(2)$. As the simplest generalization of this, Pauli replaced the circle by the surface of a 2-dimensional sphere S_2. Thus he chose $n = 3$ and $R \in SO(3)$ (to be lifted to $SU(2)$ when applied to spinors).

The next step was to satisfy (157) by making the assumption that

$$g^a_\mu = f^a_{\mu b} y^b \qquad \partial_a g_{\mu\nu} = 0. \tag{166}$$

This is stronger than (157) for general y but is stronger only by a constant term for $y = 0$.

What Pauli evidently considered to be his main result was the derivation of a quantity that would vanish if and only if the connection $\Gamma^a_{\mu b}$ could be transformed to zero. This was exactly the Riemann tensor (164) and he called it the field strength.

He also applied the formalism to spinors, in much the same way as Klein had done in 1938, and thus obtained a Dirac equation of the form

$$(\gamma^\mu D_\mu + m)\psi = 0, \tag{167}$$

where D_μ is the operator in (156). He noted that the coupling was necessarily vectorlike, in contrast to the other proposals being made at the time, but pointed out that some Lorentz scalar but matrix-valued terms might be added to the Dirac operator in order to provide other types of interaction. As examples he quoted the original Pais term $\sigma.K$ and suggested as other possibilities $i\sigma^a \frac{\partial}{\partial y^a}$ and $\gamma_5\sigma^a \frac{\partial}{\partial y^a}$, where the σ's are the Pauli matrices. He computed the mass spectrum for the spinors in these two cases but considered it unsatisfactory because of a twofold degeneracy of the states. So far as is known, Pauli did not pursue the problem any farther and did not have his notes typed, much less sent for publication.

Meson–Nucleon Interaction and Differential Geometry

by W. Pauli in Zurich

Letters to A. Pais, July, December, 1953

Meson–Nucleon Interaction and Differential Geometry

Written down July 22–25, 1953, in order to see how it looks.

Split a 6-dimensional space into a (4 + 2)-dimensional one. Consider the 2-dimensional space as a spherical surface. Introduce on it three Cartesian coordinates with the restriction

$$\Omega_1^2 + \Omega_2^2 + \Omega_3^2 = 1 \tag{1}$$

The 4-dimensional space-time, denoted by x^i or briefly x, will be left unchanged. We consider now the *group* (linear)

$$\Omega'^A = C_B^A(x)\Omega^B \tag{I}$$

(fundamental for all that follows), with x-dependent coefficients. As orthogonality conditions are not elegant, I introduce coefficients $g_{AB}(x)$ possibly depending on x (but independent of the Ω) and generalize (1) to

$$g_{AB}(x)\Omega^A\Omega^B = 1, \quad (A, B = 1, 2, 3), \tag{1a}$$

(although one can, of course, always transform the g_{AB} into δ_{AB} for all x by a suitable transformation of form (I)).

Remark. In the case of a *circle* ($A, B = 1, 2$ and my Ω^1, Ω^2 correspond to $\cos x_5$ and $\sin x_5$ of Kaluza, so that the gauge transformation $x_5' = x_5 + f(x^1, \ldots, x^4)$ is indeed of the form (I). The group (I) seems to me therefore the *natural generalization of the gauge group* in the case of a two dimensional surface. (Whether or not orthogonality conditions are included in (I) is a matter of convenience. I decided not to apply them).

The usual rules for the transformation of covariant vectors are

$$f_A' = \frac{\partial \Omega^B}{\partial \Omega'^A} f_B \tag{2a}$$

$$f_i' = f_i + \frac{\partial \Omega^B}{\partial x^i} f_B, \tag{2b}$$

taking $\frac{\partial \Omega^B}{\partial x^i}$ for fixed Ω', and considering only transformations with $x' \equiv x$. Correspondingly for a symmetrical covariant tensor (transforms like product of two vectors)

$$g'_{AB} = \frac{\partial \Omega^C}{\partial \Omega'^A} \frac{\partial \Omega^D}{\partial \Omega'^B} g_{CD}, \tag{3a}$$

$$g'_{Ai} = \frac{\partial \Omega^B}{\partial \Omega'^A} \left(g_{Bi} + \frac{\partial \Omega^C}{\partial x^i} g_{BC} \right), \tag{3b}$$

$$g'_{ik} = g_{ik} + \frac{\partial \Omega^B}{\partial x^i} g_{Bk} + \frac{\partial \Omega^B}{\partial x^k} g_{Bi} + \frac{\partial \Omega^B}{\partial x^i} \frac{\partial \Omega^C}{\partial x^k} g_{BC}. \tag{3c}$$

And for contravariant vectors

$$f'^A = \frac{\partial \Omega'^A}{\partial \Omega^B} f^B + \left(\frac{\partial \Omega'^A}{\partial x^k} \right)_{\Omega \text{ fixed}} f^k, \tag{4a}$$

$$f'^i = f^i. \tag{4b}$$

The main "*rule of the game*" for the group (I) is to eliminate the rules (2b), (3c) and (4a) with the help of a given tensor g_{AB} and g_{Ai} by defining the *underlining* of small indices below and capital indices above (for the other two possibilities the underlining is superfluous anyhow), so that the vectors (or tensors) with underlined indices transform under group (I) entirely disregarding the x-space

$$f'_{\underline{i}} = f_{\underline{i}}, \quad f'^{\underline{A}} = \frac{\partial \Omega'^A}{\partial \Omega^B} f^{\underline{B}}. \tag{5}$$

Define $g^{\underline{A}\underline{B}}$ in the usual way, being given g_{AB}, by

$$g_{AC} g^{\underline{B}\underline{C}} = \delta_A^{\ B}.$$

Use $g^{\underline{A}\underline{B}}$ to raise indices, for instance $g_i^{\ \underline{A}} = g_{iB} g^{\underline{A}\underline{B}}$, and define

$$\begin{aligned} f_{\underline{i}} &= f_i - g_i^{\ \underline{A}} f_A, \\ f^{\underline{A}} &= f^A + g_k^{\ \underline{A}} f^k. \end{aligned} \tag{6}$$

Then the transformation law (5) actually holds. The rules for underlining tensors follow immediately from this, as tensors always behave like products of vectors. For the g-tensor itself, for instance, there follows

$$g_{i\underline{A}} = 0, \qquad g_{ik} = g_{\underline{i}\,\underline{k}} + g_{iA} g_k{}^A, \tag{7a}$$

$$g^{i\underline{A}} = 0, \qquad g^{AB} = g^{\underline{A}\,\underline{B}} + g^{ik} g_i{}^A g_k{}^B; \tag{7b}$$

but

$$g^{iA} = -g^{ik} g^A_{\ k} = -g^{ik} g^{AB} g_{Bk}, \tag{7c}$$

$$f^i = g^{ik} f_k + g^{iA} f_A = g^{ik} f_{\underline{k}},$$

$$f_i = g_{ik} f^k + g_{iA} f^A, \quad f_{\underline{i}} = g_{\underline{i}\,\underline{k}} f^k,$$

$$f_A = g_{AB} f^B + g_{Ak} f^k = g_{AB} f^{\underline{B}}, \text{ etc.}$$

According to the rule (3*b*) it is natural to assume that g_{Ai} *depends linearly on the* Ω

$$g_{Ai} = f_{AB,i}(x)\,\Omega^B \tag{II}$$

since g_{AB} is independent of Ω. The assumption (II) enables us to transform away (nullify) the g_{Ai} *locally* by a suitable transformation (I), the second term in the bracket of (3*b*) depending linearly on the Ω.

We define now accordingly the Dirac-matrices by

$$\frac{1}{2}(\gamma_i\gamma_k + \gamma_k\gamma_i) = g_{ik}$$

$$\frac{1}{2}(\gamma_i\gamma_A + \gamma_A\gamma_i) = g_{iA}$$

$$\frac{1}{2}(\gamma_A\gamma_B + \gamma_B\gamma_A) = g_{AB}.$$

Hence with

$$\gamma_{\underline{i}} = \gamma_i - g_i^{\ A}\gamma_A = g_{\underline{i}\,\underline{k}}\gamma^k,$$

$$\gamma^{\underline{A}} = \gamma^A + g_k^{\ A}\gamma^k = g^{\underline{A}\,\underline{B}}\gamma_B,$$

$$\frac{1}{2}(\gamma_{\underline{i}}\gamma_{\underline{k}} + \gamma_{\underline{k}}\gamma_{\underline{i}}) = g_{\underline{i}\,\underline{k}}$$

$$\frac{1}{2}(\gamma_{\underline{i}}\gamma_A + \gamma_A\gamma_{\underline{i}}) = 0$$

$$\frac{1}{2}(\gamma^i\gamma^k + \gamma^k\gamma^i) = g^{ik}$$

$$\frac{1}{2}(\gamma^i\gamma^{\underline{A}} + \gamma^{\underline{A}}\gamma^i) = 0$$

$$\frac{1}{2}(\gamma^{\underline{A}}\gamma^{\underline{B}} + \gamma^{\underline{B}}\gamma^{\underline{A}}) = g^{\underline{A}\,\underline{B}}.$$

For our purpose (disregarding gravitation) it is sufficient to identify $g_{\underline{i}\,\underline{k}}$ and g^{ik} with the diag(1, 1, 1, −1) of special relativity; similarly with γ^i and $\gamma_{\underline{i}}$.

One can identify $\gamma^{\underline{A}}$ with $\gamma^5\tau^A$, τ^A being the isotopic spin commuting with the $\gamma_{\underline{i}}$. I omit here the usual discussion of Hermiticity.

The Dirac equation becomes (see (II))

$$\gamma^k\psi_{;\,k} + \gamma^A\frac{\partial\psi}{\partial\Omega^A} + M\psi = 0$$

or

$$\gamma^k\psi_{;\,k} + (\gamma^{\underline{A}} - \gamma^k f^{\underline{A}}_{B,k}(x)\Omega^B)\frac{\partial\psi}{\partial\Omega^A} + M\psi = 0 \tag{7}$$

It is interesting that a vector-coupling results and not the Cornell-factory-product (for which I propose the name "hot dog potential"). But I do not see another way to be faithful to the group (I). And if I am supposed to believe in your Ω-space, I have to believe in the group (I) too. And a scalar field I can transform away by (I), except as a factor of the mass-term.

Remarks. 1. We define

$$\psi_{;\,k} = \frac{\partial\psi}{\partial x^k} + \Lambda_k\psi$$

with a suitable Λ_k. In the coordinate system where $g_{AB} = \delta_{AB}$ one has $\Lambda_k = 0$.

2. The homogeneous character of the Ω (see ($1a$) above) can be taken care of by adding a suitably chosen $\lambda g_{AB}\Omega^B$ to every $\frac{\partial}{\partial\Omega^A}$. One can fix λ by claiming $\Omega^A\left(\frac{\partial\psi}{\partial\Omega^A}\right) = 0$. λ is arbitrary and possibly Ω-dependent. The homogeneous coordinates are very convenient for practical computations as the spherical harmonics are polynomials in the Ω.

For the case of the circle ($A, B = 1, 2$ only) one can try to write the Dirac equation in an electromagnetic field in this way (Kaluza, Klein). But then only very particular solutions of (7) are interpreted (and therefore I could not accept it). By developing ψ in Fourier series on the circle, one obtains a quantum number n (which is a *component* of your "orbital-momentum quantum number" of the Ω-space, which in your case gives rise to S, P, D, ... terms). Then only the lowest values of n ($n = 0$ or 1) could be interpreted in the Kaluza-Klein fashion.

The constant term $M\psi$ may be superfluous (?). I am looking forward to getting your paper. Would you like to assume a field $f(x)\psi$ instead of $M\psi$? Or $(F_B(x)\Omega^B)\psi$?

And now comes my main result. What are the necessary and sufficient conditions, that g_{Ai} (or equivalently $f_{AB,i}(x)$) can be transformed away in the whole x-space?

Answer. There exists a *tensor*

$$F^{A}_{\underline{ik}} = f^{A}_{B\underline{ik}}\Omega^B \equiv \frac{\partial g^{A}_{k}}{\partial x^i} - \frac{\partial g^{A}_{i}}{\partial x^k} + \frac{\partial g^{A}_{i}}{\partial \Omega^B} g^{B}_{k} - \frac{\partial g^{A}_{k}}{\partial \Omega^B} g^{B}_{i} \tag{8}$$

or equivalently

$$f^{A}_{B\underline{ik}}(x) \equiv \frac{\partial f^{A}_{B,k}}{\partial x^i} - \frac{\partial f^{A}_{B,i}}{\partial x^k} + f^{A}_{C,i} f^{C}_{B,k} - f^{A}_{C,k} f^{C}_{B,i} \tag{8a}$$

This is the *true* physical field, the analogue of the field-strengths. If and only if they vanish, the field $f^{A}_{B,i}$ can be made identically zero for all x by a transformation (I).

I have checked the tensor-character of (8) in different ways.

Mathematical Appendix

(written December 1953).

Ich möchte hier einige Präzisierungen, Verbesserungen und Ergänzungen zusammenstellen zu dem, was ich in Sommer geschrieben habe.*

1. Consider the linear transformations

$$\begin{aligned} \Omega'^A &= a^{A}_{B}(x)\Omega^B \\ \text{inverse } \Omega^B &= \overline{a}^{B}_{A}(x)\Omega'^A \\ a^{A}_{B}\overline{a}^{B}_{C} &= \overline{a}^{A}_{B} a^{B}_{C} = \delta^{A}_{C}. \end{aligned} \tag{I}$$

*I would like to make what I wrote last Summer more precise, and to add some improvements and extensions. This was the only German sentence in the (otherwise English) Pauli-Pais letters.

Define vectors A^ρ or B_ρ ($\rho = i$ or A) which transform according to

$$A'^i = \frac{\partial x'^i}{\partial x^k} A^k$$

$$\begin{aligned} A'^A =& \frac{\partial \Omega'^A}{\partial \Omega^B} A^B + \frac{\partial \Omega' A}{\partial x^k} A^k \\ =& a^A{}_B(x) A^B + \frac{\partial a^A{}_B}{\partial x^k} \Omega^B A^k \end{aligned}$$

and

$$\begin{aligned} B'_i =& \frac{\partial x^k}{\partial x'^i} B_k + \frac{\partial \Omega^A}{\partial x'^i} B_A \\ =& \frac{\partial x^k}{\partial x'^i} B_k + \frac{\partial \overline{a}^C{}_A}{\partial x'^i} \Omega'^A B_C \end{aligned}$$

$$B'_A = \frac{\partial \Omega^C}{\partial \Omega'^A} B_C = \overline{a}^C{}_A B_C.$$

Assumption 1.

Existence of a metric tensor $g_{\rho\sigma} = (g_{ik}, g_{iA}, g_{AB})$ which transforms like $B_\rho B_\sigma$. In particular,

$$g'_{iA} = \frac{\partial \Omega^B}{\partial \Omega'^A} \left(g_{iB} + \left(\frac{\partial \Omega^C}{\partial x^i} \right)_{\Omega'} g_{BC} \right). \qquad (*)$$

We use $g_{AB}(x)$ to fix the homogeneity condition

$$g_{AB} \Omega^A \Omega^B = 1. \qquad (1)$$

We use the normalized cofactors of the g_{AB}, defined by

$$g^{\underline{A}\,\underline{C}} g_{CB} = \delta^{\underline{A}}_{C}, \qquad (2)$$

to construct from X^ρ and Y_ρ vectors with *underlined* indices. Let

$$g_{i}^{\underline{A}} = g^{\underline{A}\,\underline{B}} g_{Bi} \qquad (3)$$

and

$$X^{\underline{A}} = X^A + g_i{}^A X^i, \quad Y_{\underline{i}} = Y_i - g_i{}^A Y_A$$

Then

$$\left.\begin{aligned} X'^A &= a^A_{\ B}(x) X^{\underline{B}} \\ Y'_i &= \tfrac{\partial x^k}{\partial x'^i} Y_k \end{aligned}\right] \quad \text{without extra terms.} \tag{4}$$

$$X^i Y_{\underline{i}} + X^{\underline{A}} Y_A = X^i Y_i + X^A Y_A.$$

Analogously we define

$$g_{\underline{i}\,\underline{k}} = g_{ik} - g_i^{\ A} g_{Ak} - g_k^{\ A} g_{iA},$$

which is invariant under pure Ω-transformations with $x'^i = x^i$.

Assumption 2.

g_{AB} is independent of Ω, and g_{Ai} is linear and homogeneous in Ω. This means that in each point of x-space (for all Ω) the g_{Ai} can be transformed to 0 and the g_{AB} to δ_{AB}. Thus

$$g_{iA} = f_{i,AB}(x)\Omega^B \tag{6}$$

with $f_{i,AB}$ independent of Ω. By (*)

$$f'_{i,AB} = \overline{a}^C_{\ A}\left(\overline{a}^D_{\ B}\, f_{i,CD} + \frac{\partial \overline{a}^D_{\ B}}{\partial x^i} g_{CD}\right). \tag{7}$$

This is the generalization of the "gauge-group".

Construction of tensors by differentiation (this is new). We find that

$$f_{i,AB} + f_{i,BA} - \frac{\partial g_{AB}}{\partial x^i} = 2\{\underline{i}, AB\} \tag{8}$$

is a tensor with underlined i-index. One might try to introduce a pseudoscalar field by means of this tensor, but it seems to me artificial.

In what follows I now make the

Assumption 3.

$$\{\underline{i}, AB\} = 0, \ \text{or} \ f_{i,AB} = \frac{1}{2}\frac{\partial g_{AB}}{\partial x^i} + \phi_{i,[AB]}. \tag{9}$$

Here ϕ is antisymmetric in A, B.

This means: we can make $g_{AB} = \delta_{AB}$ in the whole x-space, and also $g_{Ai} = 0$ at any particular point.

Rule. The operator

$$\nabla_{\underline{k}} = \frac{\partial}{\partial x^k} - g^{\underline{A}\underline{B}} f_{k,BC}\Omega^C \frac{\partial}{\partial \Omega^A}, \tag{10}$$

applied to any scalar depending on x and Ω, gives a vector with underlined lower index k. (Special case of the formula for $Y_{\underline{i}}$).

Rule. Use the usual 3-index symbol defined by

$$\begin{aligned}\left\{\begin{matrix}\underline{A}\\ B\,i\end{matrix}\right\} &= g^{\underline{A}\underline{C}}\frac{1}{2}\left(\frac{\partial g_{BC}}{\partial x^i} + \frac{\partial g_{Ci}}{\partial \Omega^B} - \frac{\partial g_{Bi}}{\partial \Omega^C}\right)\\ &= g^{\underline{A}\underline{C}}\left(\frac{1}{2}\frac{\partial g_{BC}}{\partial x^i} - \phi_{i,[BC]}\right) \text{ by (9).}\end{aligned} \tag{11}$$

Then

$$X^{\underline{A}}_{;\underline{k}} = \frac{\partial X^{\underline{A}}}{\partial x_k} + \left\{\begin{matrix}\underline{A}\\ B\,k\end{matrix}\right\} X^{\underline{B}} \tag{12}$$

is a vector with underlined index k.

Rule. The "field-strengths" follow from (10) as follows

$$F^{\underline{A}}_{,\underline{i}\,\underline{k}} = \nabla_{\underline{i}}\, g^{\underline{A}}_{\underline{k}} - \nabla_{\underline{k}}\, g^{\underline{A}}_{\underline{i}}.$$

Or with

$$g_{AB} F^{\underline{B}}_{,\underline{i}\,\underline{k}} = f_{AB,\underline{i}\,\underline{k}}\Omega^B,$$

$$\begin{aligned}f_{AB,\underline{i}\,\underline{k}} = &\frac{\partial f_{AB,k}}{\partial x^i} - \frac{\partial f_{AB,i}}{\partial x^k} - g^{CD}\left(\frac{\partial g_{AD}}{\partial x^i} f_{k,CB} - \frac{\partial g_{AD}}{\partial x^k} f_{i,CB}\right)\\ &- g^{CD}\left(f_{k,AC} f_{i,DB} - f_{i,AC} f_{k,DB}\right).\end{aligned} \tag{13}$$

The field-strengths are antisymmetric in i, k. By (9) they are also antisymmetric in A, B. Always assuming (9), *the vanishing of the field-strengths is necessary and sufficient for the $f_{i,AB}$ in the whole space to be transformable to zero.*

Dirac matrices and Dirac equation.

For simplicity we assume Special Relativity theory, so $g_{ik} = g^o_{ik} = \mathrm{diag}(1, 1, 1, -1)$ and $g^{ik} = g^o_{ik}$. Also

$$\gamma^i\gamma^k + \gamma^k\gamma^i = 2g^{oik}, \tag{14}$$

i.e. the γ^i are ordinary Dirac matrices independent of x and Ω. Likewise

$$\gamma^{\underline{A}}\gamma^{\underline{B}} + \gamma^{\underline{B}}\gamma^{\underline{A}} = 2g^{\underline{A}\underline{B}}, \tag{15}$$

$\gamma^{\underline{A}}$ independent of Ω. By (11), (12) we may define

$$\frac{\partial\gamma^{\underline{A}}}{\partial x^i} + \left\{ \begin{matrix} A \\ B\,i \end{matrix} \right\} \gamma^{\underline{B}} + \Lambda_{\underline{i}}\gamma^{\underline{A}} - \gamma^{\underline{A}}\Lambda_{\underline{i}} = 0, \tag{16}$$

$$\Lambda_{\underline{i}}\gamma^k - \gamma^k\Lambda_{\underline{i}} = 0,$$

the existence of the $\Lambda_{\underline{i}}$ following from (15) according to Schrödinger. Using (10), we assume now

$$\gamma^k(\nabla_{\underline{k}} + \Lambda_{\underline{k}})\psi + \mu\Sigma + m\psi = 0. \tag{17}$$

m = nucleon mass, μ = your Λ^{-1}, Σ is something invariant. For the adjoint

$$\nabla_{\underline{k}}\overline{\psi}\gamma^k - \overline{\psi}\Lambda_{\underline{k}}\gamma^k + \mu\overline{\Sigma} - \overline{\psi}m = 0. \tag{17a}$$

From the point of view of the group (I), the "something" Σ is quite unnecessary, it could be zero. We could make various assumptions about it. Your assumption (in your notation)

$$\Sigma = \tau \cdot K$$

is *consistent* with the group (I), but is also arbitrary.

I discuss here other possible choices for Σ, using for brevity the coordinate system in which $g_{AB} = \delta_{AB}$, so that (I) reduces to the x-dependent *orthogonal* group.

1.) Put

$$\Sigma = i\tau^A \frac{\partial}{\partial\Omega^A}, \qquad \overline{\Sigma} = -i\tau^A \frac{\partial}{\partial\Omega^A},$$

τ^A commuting with γ^k. The factor i is necessary, so that $\overline{\psi} = \psi^*\gamma^4$ and

$$\frac{\partial}{\partial x^k}(\overline{\psi}\gamma^k\psi) + \frac{\partial}{\partial\Omega^A}(\overline{\psi}\gamma^A\psi) = 0.$$

[In general coordinates we would have instead of this $\frac{1}{\sqrt{g}}\frac{\partial}{\partial x^k}(\overline{\psi}\sqrt{g}\gamma^k\psi)$ with $g = \det||g_{AB}||$.]

The homogeneity condition, here $\sum_A \Omega_A^2 = 1$, is to be taken into account in defining $\frac{\partial}{\partial\Omega^A}$, so that $\frac{\partial}{\partial\Omega^A}$ is only the component of "grad" *tangent* to the sphere, or normal to the vector $\vec{\Omega}$. I found the equations in my old paper (*Helv. Phys. Acta* **12**, 147, 1939) useful for discovering the eigenvalues of Σ.

The result is: Σ has two eigenvalues

$$+ \text{ and } - \sqrt{(j+\tfrac{1}{2})^2 - 1}, \quad j = \tfrac{1}{2}, \tfrac{3}{2}, \ldots \tag{18}$$

and each of them is simple. In the field-free case $f_{i,AB} = 0$, then (17) gives for the "Mass Operator" $\gamma^k \frac{\partial}{\partial x^k}$ the eigenvalues

$$\text{“}M\text{”} = m \pm \mu\sqrt{(j+\tfrac{1}{2})^2 - 1} \tag{18$_1$}$$

whose *absolute* values are the rest-masses.

2.) Put

$$\Sigma = \gamma^5 \tau^A \frac{\partial}{\partial\Omega^A},$$

this time without i, so Σ *anti*commutes with γ^k. This choice is reasonable, if the group (I) is enlarged. Then

$$[\text{“Mass-operator”} \mp i\sqrt{(j+\tfrac{1}{2})^2 - 1}\,\gamma^5 + m]\psi = 0$$

and the eigenvalues of "Mass-operator" are

$$\text{“}M\text{”} = \pm\sqrt{m^2 + \mu^2[(j+\tfrac{1}{2})^2 - 1]}. \tag{18$_2$}$$

This last result is due to Touschek (Rome), who also made the following comments.

Σ, as written down here, is *not* invariant under reflection $\Omega' = -\Omega$ of the Ω-space (Case 2 only). Therefore there is no longer invariance under reflection ($x_4' = x_4$, $x_\alpha' = -x_\alpha$ for $\alpha = 1, 2, 3$) in the x-space. This is ugly, and also means that the Parity in Ω-space is no longer a good quantum number (it does not commute with the Hamiltonian). To overcome this difficulty, we must replace the τ^A in this expression for Σ by 4-row matrices $\begin{pmatrix} \tau^A & \\ & -\tau^A \end{pmatrix}$ or something equivalent. Then we can put $k^2 = (j+\frac{1}{2})^2$, $k = \pm 1, \pm 2, \ldots$. The parity is $\frac{k}{j+\frac{1}{2}}$ and *each* eigenvalue

$$+\sqrt{k^2 - 1} \text{ and } -\sqrt{k^2 - 1}$$

of Σ [in case 2 multiplied by $i\gamma^5$] is doubly degenerate because of the two alternative signs of k.

So this leads to some rather unphysical "shadow particles".

8

The Yang-Mills Theory, 1953–54

Even while Pauli was writing his first letter to Pais, the problem of constructing a non-abelian gauge theory was being brought to a successful conclusion by Yang and Mills. Motivated by the desire to make the the isotopic spin symmetry of the strong interactions local, they invented the $SU(2)$ gauge theory that has become synonymous with their names. The original papers are included here, and the Yang-Mills theory is so well-known and the presentation in these papers was so clear that they require no discussion. Some historical comments might not be out of place, however.

As mentioned in the Introduction Yang had been considering the problem since 1949. To quote from his Selected Papers [3]:

> While a graduate student in Kunming and in Chicago, I had thoroughly studied Pauli's review articles on field theory. I was very much impressed with the idea that charge conservation was related to the invariance of the theory under phase changes, an idea, I later found out, due originally to Weyl. I was even more impressed by the fact that gauge-invariance *determined* all the electromagnetic interactions. While in Chicago I tried to generalize this to istopic spin interactions by the procedure later written up in [4] [i.e., the accompanying Phys. Rev. **96** (1954) 191 article] equations (1) and (2). Starting from these it was easy to get equation (3). Then I tried to define the field strengths $F_{\mu\nu}$ by $F_{\mu\nu} = \partial_\mu B_\nu - \partial_\nu B_\mu$ which was a natural generalization of electromagnetism. This led to a mess, and I had to give up. But the basic idea remained attractive, and I came back to it several times in the next few years, always getting stuck at the same point .... As more and more mesons were discovered and all kinds of interactions were being considered, the necessity to have a *principle* for writing down interactions became more obvious to me. So while at Brookhaven [in the summer of 1953] I returned once more to the idea of generalizing gauge invariance. My office mate was R. L. Mills who was about to finish his Ph.D.... We worked on the problem and eventually produced [4]. We also wrote an Abstract for the April 1954 meeting of the AMS in Washington, which became [5]. Different motivations were emphasized in the two papers. The formal aspect of the work did not take long and was essentially finished by February 1954. But we found that we were unable to conclude what the mass of the gauge-particles should be. We

> toyed with the dimensional argument that, for a pure gauge theory, there is no quantity with the dimension of mass to start with, and therefore a gauge particle must be massless. But we quickly rejected this line of reasoning.

It is interesting to consider this account from a present-day perspective. At first glance, it is a little surprising that, having successfully found the expression for the covariant derivatives D_μ, Yang had such difficulty in finding the expression for the covariant field strengths, since these are the simple functions $F_{\mu\nu} = [D_\mu, D_\nu]$ of the D_μ's. However, the expression of the field strength as a commutator was not common in electromagnetic theory at the time because the direct expression as a curl was so simple, so the commutator expression was not at Yang's disposal. The foresight that warned Yang and Mills that the gauge-field mass problem would not have a simple solution is also interesting. With hindsight, we now know that for the weak interactions this problem is solved only by spontaneous symmetry breaking and that for the strong interactions it remains a deep problem that is intimately connected with asymptotic freedom and confinement.

The question of the gauge-field mass problem was raised by Pauli when Yang was invited to present the Yang-Mills results at the Princeton Institute in February 1954. As Yang [3] relates:

> Pauli was spending the year in Princeton, and was deeply interested in symmetries and interactions. . . . Soon after my seminar began, when I had written on the blackboard,
>
> $$(\partial_\mu - ieB_\mu)\psi,$$
>
> Pauli asked, "What is the mass of this field B_μ?" I said we did not know. Then I resumed my presentation but soon Pauli asked the same question again. I said something to the effect that it was a very complicated problem, we had worked on it and had come to no definite conclusions. I still remember his repartee: "That is not sufficient excuse". I was so taken aback that I decided, after a few moments' hesitation, to sit down. There was general embarrassment. Finally Oppenheimer said, "We should let Frank proceed". I then resumed and Pauli did not ask any more questions during the seminar.

Thus Pauli also was aware of the non-triviality of the mass problem. But some other interesting information is revealed by the following sequel to the episode:

> I do not remember what happened at the end of the seminar. But next day I found the following message:
>
> February 24, Dear Yang, I regret that you made it almost impossible for me to talk to you after the seminar. All good wishes. Sincerely yours, W. Pauli.

> I went to talk to Pauli. He said I should look up a paper by E. Schrödinger... it was a discussion of space-time dependent representations of the γ_μ matrices for a Dirac electron in a gravitational field. Equations in it were, on the one hand, related to equations in Riemannian geometry and, on the other, similar to the equations that Mills and I were working on. But it was many years later when I understood that these were all different cases of the mathematical theory of connections on fibre bundles.

This statement is interesting because it shows that when Yang and Mills were constructing their theory they had no idea that it might be related to gravitation. This had been stated by Yang on other occasions also. For example, in the *Proceedings of the Weyl Centenary Meeting* [6], referring to Weyl's 1929 statement, "Since gauge-invariance involves an arbitrary function λ it has the character of general relativity and naturally can only be understood in that context," he says:

> The above quotation also contains something that is very revealing, namely, his [Weyl's] strong association of gauge-invariance with general relativity.... Twenty years later, when Mills and I worked on non-abelian gauge-fields, our motivation was completely divorced from general relativity and we did not appreciate that gauge fields and general relativity are somehow related. Only in the late 1960's did I recognize the structural similarity mathematically of non-abelian gauge-fields with general relativity and understand that they were both connections mathematically.

From the historical point of view, it is also interesting to know that Yang and Weyl overlapped for some years at Princeton. But it is astonishing to find that, although they met from time to time, they did not discuss gauge theory, or indeed physics or mathematics of any kind. To quote Yang [6] once more:

> I had met Weyl in 1949 when I went to the IAS as a young 'member'. I saw him from time to time in the next years, 1949-55. He was very approachable, but I do not remember having discussed physics or mathematics with him at any time. Neither Pauli nor Oppenheimer ever mentioned it. I suspect they also did not tell Weyl of the 1954 papers of Mills' and mine. Had they done that, or had Weyl somehow come across our paper, I imagine that he would have been pleased and excited, for we had put together two things that were very close to his heart: gauge invariance and non-abelian Lie groups.

Isotopic Spin Conservation and a Generalized Gauge Invariance*

by C. N. Yang and R. Mills in Upton, N.Y.

Phys. Rev. 95 (1954) 631

The conservation of isotopic spin points to the existence of a fundamental invariance law similar to the conservation of electric charge. In the latter case, the electric charge serves as a source of electromagnetic field; an important concept in this case is that a gauge invariance which is closely connected with (1) the equation of motion of the electromagnetic field, (2) the existence of a current density, and (3) the possible interactions between a charged field and the electromagnetic field. We have tried to generalize this concept of gauge invariance to apply to isotopic spin conservation. It turns out that a very natural generalization is possible. The field that plays the role of the electromagnetic field is here a vector field that satisfies a nonlinear equation even in the absence of other fields. (This is because unlike the electromagnetic field this field has an isotopic spin and consequently acts as a source of itself.) The existence of a current density is automatic, and the interaction of this field with any fields of arbitrary isotopic spin is of a definite form (except for possible terms similar to the anomalous magnetic moment interaction terms in electrodynamics.)

*Work performed under the auspices of the U. S. Atomic Energy Commission.

Conservation of Isotopic Spin and Isotopic Gauge Invariance*

by C. N. Yang and R. L. Mills in Upton, N.Y.

Phys. Rev. 96 (1954) 191

It is pointed out that the usual principle of invariance under isotopic spin rotation is not consistent with the concept of localized fields. The possibility is explored of having invariance under local isotopic spin rotations. This leads to formulating a principle of isotopic gauge invariance and the existence of a $\mathbf{b}$ field which has the same relation to the isotopic spin that the electromagnetic field has to the electric charge. The $\mathbf{b}$ field satisfies nonlinear differential equations. The quanta of the $\mathbf{b}$ field are particles with spin unity, isotopic spin unity, and electric charge $\pm e$ or zero.

Introduction

The conservation of isotopic spin is a much discussed concept in recent years. Historically an isotopic spin parameter was first introduced by Heisenberg[1] in 1932 to describe the two charge states (namely neutron and proton) of a nucleon. The idea that the neutron and proton correspond to two states of the same particle was suggested at that time by the fact that their masses are nearly equal, and that the light stable even nuclei contain equal numbers of them. Then in 1937 Breit, Condon, and Present pointed out the approximate equality of $p-p$ and $n-p$ interactions in the ^1S state.[2] It seemed natural to assume that this equality holds also in the other states available to both the $n-p$ and $p-p$ systems. Under such an assumption one arrives at the concept of a total isotopic spin[3] which is conserved in nucleon-nucleon interactions. Experiments in recent years[4] on the energy levels of light nuclei strongly suggest that this assumption

*Work performed under the auspices of the U. S. Atomic Energy Commission.

[1] W. Heisenberg, Z. Physik **77**, 1 (1932).

[2] Breit, Condon, and Present, Phys. Rev. **50**, 825 (1936). J. Schwinger pointed out that the small difference may be attributed to magnetic interactions [Phys. Rev. **78**, 135 (1950)].

[3] The total isotopic spin **T** was first introduced by E. Wigner, Phys. Rev. **51**, 106 (1937); B. Cassen and E. U. Condon, Phys. Rev. **50**, 846 (1936).

[4] T. Lauritsen, Ann. Rev. Nuclear Sci. **1**, 67 (1952); D. R. Inglis, Revs. Modern Phys. **25**, 390 (1953).

is indeed correct. An implication of this is that all strong interactions such as the pion-nucleon interaction, must also satisfy the same conservation law. This and the knowledge that there are three charge states of the pion, and that pions can be coupled to the nucleon field *singly*, lead to the conclusion that pions have isotopic spin unity. A direct verification of this conclusion was found in the experiment of Hildebrand[5] which compares the differential cross section of the process $n + p \rightarrow \pi^0 + d$ with that of the previously measured process $p + p \rightarrow \pi^+ + d$.

The conservation of isotopic spin is identical with the requirement of invariance of all interactions under isotopic spin rotation. This means that when electromagnetic interactions can be neglected, as we shall hereafter assume to be the case, the orientation of the isotopic spin is of no physical significance. The differentiation between a neutron and a proton is then a purely arbitrary process. As usually conceived, however, this arbitrariness is subject to the following limitation: once one chooses what to call a proton, what a neutron, at one space-time point, one is then not free to make any choices at other space-time points.

It seems that this is not consistent with the localized field concept that underlies the usual physical theories. In the present paper we wish to explore the possibility of requiring all interactions to be invariant under *independent* rotations of the isotopic spin at all space-time points, so that the relative orientation of the isotopic spin at two space-time points becomes a physically meaningless quantity (the electromagnetic field being neglected).

We wish to point out that an entirely similar situation arises with respect to the ordinary gauge invariance of a charged field which is described by a complex wave function ψ. A change of gauge[6] means a change of phase factor $\psi \rightarrow \psi'$, $\psi' = (\exp i\alpha)\psi$, a change that is devoid of any physical consequences. Since ψ may depend on x, y, z, and t, the relative phase factor of ψ at two different space-time points is therefore completely arbitrary. In other words, the arbitrariness in choosing the phase factor is local in character.

We define *isotopic gauge* as an arbitrary way of choosing the orientation of the isotopic spin axes at all space-time points, in analogy with the electromagnetic gauge which represents an arbitrary way of choosing the complex phase factor of a charged field at all space-time points. We then propose that all physical processes (not involving the electromagnetic field) be invariant under an isotopic gauge transformation, $\psi \rightarrow \psi'$, $\psi' = S^{-1}\psi$, where S represents a space-time dependent isotopic spin rotation.

To preserve invariance one notices that in electrodynamics it is necessary to counteract the variation of α with x, y, z, and t by introducing the electromag-

[5] R. H. Hildebrand, Phys. Rev. **89**, 1090 (1953).
[6] W. Pauli, Revs. Modern Phys. **13**, 203 (1941).

netic field A_μ which changes under a gauge transformation as

$$A'_\mu = A_\mu + \frac{1}{e}\frac{\partial \alpha}{\partial x_\mu}.$$

In an entirely similar manner we introduce a B field in the case of the isotopic gauge transformation to counteract the dependence of S on x, y, z, and t. It will be seen that this natural generalization allows for very little arbitrariness. The field equations satisfied by the twelve independent components of the B field, which we shall call the **b** field, and their interaction with any field having an isotopic spin are essentially fixed, in much the same way that the free electromagnetic field and its interaction with charged fields are essentially determined by the requirement of gauge invariance.

In the following two sections we put down the mathematical formulation of the idea of isotopic gauge invariance discussed above. We then proceed to the quantization of the field equations for the **b** field. In the last section the properties of the quanta of the **b** field are discussed.

Isotopic Gauge Transformation

Let ψ be a two-component wave function describing a field with isotopic spin $\frac{1}{2}$. Under an isotopic gauge transformation it transforms by

$$\psi = S\psi', \tag{1}$$

where S is a 2×2 unitary matrix with determinant unity. In accordance with the discussion in the previous section, we require, in analogy with the electromagnetic case, that all derivatives of ψ appear in the following combination:

$$(\partial_\mu - i\epsilon B_\mu)\psi.$$

B_μ are 2×2 matrices such that[7] for $\mu = 1, 2$, and 3, B_μ is Hermitian and B_4 is anti-Hermitian. Invariance requires that

$$S(\partial_\mu - i\epsilon B'_\mu)\psi' = (\partial_\mu - i\epsilon B_\mu)\psi. \tag{2}$$

Combining (1) and (2), we obtain the isotopic gauge transformation on B_μ:

$$B'_\mu = S^{-1}B_\mu S + \frac{i}{\epsilon}S^{-1}\frac{\partial S}{\partial x_\mu}. \tag{3}$$

[7] We use the conventions $\hbar = c = 1$, and $x_4 = it$. Bold-face type refers to vectors in isotopic space, not in space-time.

The last term is similar to the gradiant term in the gauge transformation of electromagnetic potentials. In analogy to the procedure of obtaining gauge invariant field strengths in the electromagnetic case, we define now

$$F_{\mu\nu} = \frac{\partial B_\mu}{\partial x_\nu} - \frac{\partial B_\nu}{\partial x_\mu} + i\epsilon(B_\mu B_\nu - B_\nu B_\mu). \tag{4}$$

One easily shows from (3) that

$$F'_{\mu\nu} = S^{-1} F_{\mu\nu} S \tag{5}$$

under an isotopic gauge transformation.‡ Other simple functions of B than (4) do not lead to such a simple transformation property.

The above lines of thought can be applied to any field ψ with arbitrary isotopic spin. One need only use other representations S of rotations in three-dimensional space. It is reasonable to assume that different fields with the same total isotopic spin, hence belonging to the same representation S, interact with the same matrix field B_μ. (This is analogous to the fact that the electromagnetic field interacts in the same way with any charged particle, regardless of the nature of the particle. If different fields interact with different and independent B fields, there would be more conservation laws than simply the conservation of total isotopic spin.) To find a more explicit form for the B fields and to relate the B_μ's corresponding to different representations S, we proceed as follows.

Equation (3) is valid for any S and its corresponding B_μ. Now the matrix $S^{-1}\partial S/\partial x_\mu$ appearing in (3) is a linear combination of the isotopic spin "angular momentum" matrices T^i $(i = 1, 2, 3)$ corresponding to the isotopic spin of the ψ field we are considering. So B_μ itself must also contain a linear combination of the matrices T^i. But any part of B_μ in addition to this, $\bar{B}_\mu$, say, is a scalar or tensor combination of the T's, and must transform by the homogeneous part of (3), $\bar{B}'_\mu = S^{-1}\bar{B}_\mu S$. Such a field is extraneous; it was allowed by the very general form we assumed for the B field, but is irrelevant to the question of isotopic gauge. Thus the relevant part of the B field is of the form

$$B_\mu = 2\mathbf{b}_\mu \cdot \mathbf{T}. \tag{6}$$

(Bold-face letters denote three-component vectors in isotopic space.) To relate the $\mathbf{b}_\mu$'s corresponding to different representations S we now consider the product representation $S = S^{(a)}S^{(b)}$. The B field for the combination transforms,

‡*Note added in proof.* —It may appear that B_μ could be introduced as an auxiliary quantity to accomplish invariance, but need not be regarded as a field variable by itself. It is to be emphasized that such a procedure violates the principle of invariance. Every quantity that is not a pure numeral (like 2, or M, or any definite representation of the γ matrices) should be regarded as a dynamical variable, and should be varied in the Lagrangian to yield an equation of motion. Thus the quantities B_μ must be regarded as independent fields.

according to (3), by

$$B'_\mu = \left[S^{(b)}\right]^{-1}\left[S^{(a)}\right]^{-1} BS^{(a)}S^{(b)} + \frac{i}{\epsilon}\left[S^{(a)}\right]^{-1}\frac{\partial S^{(a)}}{\partial x_\mu} + \frac{i}{\epsilon}\left[S^{(b)}\right]^{-1}\frac{\partial S^{(b)}}{\partial x_\mu}.$$

But the sum of $B_\mu^{(a)}$ and $B_\mu^{(b)}$, the B fields corresponding to $S^{(a)}$ and $S^{(b)}$, transforms in exactly the same way, so that

$$B_\mu = B_\mu^{(a)} + B_\mu^{(b)}$$

(plus possible terms which transform homogenously, and hence are irrelevant and will not be included). Decomposing $S^{(a)}S^{(b)}$ into irreducible representations, we see that the twelve-component field $\mathbf{b}_\mu$ in Eq. (6) is the same for all representations.

To obtain the interaction between any field ψ of arbitrary isotopic spin with the $\mathbf{b}$ field one therefore simply replaces the gradiant of ψ by

$$(\partial_\mu - si\epsilon\mathbf{b}_\mu \cdot \mathbf{T})\psi, \tag{7}$$

where T^i $(i = 1, 2, 3)$, as defined above, are the isotopic spin "angular momentum" matrices for the field ψ.

We remark that the nine components of $\mathbf{b}_\mu$, $\mu = 1, 2, 3$ are real and the three of $\mathbf{b}_4$ are pure imaginary. The isotopic-gauge covariant field quantities $F_{\mu\nu}$ are expressible in terms of $\mathbf{b}_\mu$:

$$F_{\mu\nu} = s\mathbf{f}_{\mu\nu} \cdot \mathbf{T}, \tag{8}$$

where

$$\mathbf{f}_{\mu\nu} = \frac{\partial \mathbf{b}_\mu}{\partial x_\nu} - \frac{\partial \mathbf{b}_\nu}{\partial x_\mu} - 2\epsilon\mathbf{b}_\mu \times \mathbf{b}_\nu. \tag{9}$$

$\mathbf{f}_{\mu\nu}$ transforms like a vector under an isotopic gauge transformation. Obviously the same $\mathbf{f}_{\mu\nu}$ interact with all fields ψ irrespective of the representation S that ψ belongs to.

The corresponding transformation of $\mathbf{b}_\mu$ is cumbersome. One need, however, study only the infinitesimal isotopic gauge transformations,

$$S = 1 - 2i\mathbf{T} \cdot \delta\omega.$$

Then

$$\mathbf{b}'_\mu = \mathbf{b}_\mu + 2\mathbf{b}_\mu \times \delta\omega + \frac{1}{\epsilon}\frac{\partial}{\partial x_\mu}\delta\omega. \tag{10}$$

Field Equations

To write down the field equations for the **b** field we clearly only want to use isotopic gauge invariant quantities. In analogy with the electromagnetic case we therefore write down the following Lagrangian density:[8]

$$-\tfrac{1}{4}\mathbf{f}_{\mu\nu}\cdot\mathbf{f}_{\mu\nu}.$$

Since the inclusion of a field with isotopic spin $\frac{1}{2}$ is illustrative, and does not complicate matters very much, we shall use the following total Lagrangian density:

$$\mathcal{L}=\tfrac{1}{4}\mathbf{f}_{\mu\nu}\cdot\mathbf{f}_{\mu\nu}=\bar{\psi}\gamma_{\mu}(\partial_{\mu}-i\epsilon\tau\cdot\mathbf{b}_{\mu})\psi-m\bar{\psi}\psi. \tag{11}$$

One obtains from this the following equations of motion:

$$\begin{aligned}\partial\mathbf{f}_{\mu\nu}/\partial x_{\nu}+2\epsilon(\mathbf{b}_{\nu}\times\mathbf{f}_{\mu\nu})+\mathbf{J}_{\mu}&=0,\\ \gamma_{\mu}(\partial_{\mu}-i\epsilon\tau\cdot\mathbf{b}_{\mu})\psi+m\psi&=0,\end{aligned} \tag{12}$$

where

$$\mathbf{J}_{\mu}=i\epsilon\bar{\psi}\gamma_{\mu}\tau\psi. \tag{13}$$

The divergence of $\mathbf{J}_{\mu}$ does not vanish. Instead it can easily be shown from (13) that

$$\partial\mathbf{J}_{\mu}/\partial x_{\mu}=-2\epsilon\mathbf{b}_{\mu}\times\mathbf{J}_{\mu}. \tag{14}$$

If we define, however,

$$\mathfrak{J}_{\mu}=\mathbf{J}_{\mu}+2\epsilon\mathbf{b}_{\nu}\times\mathbf{f}\mu\nu, \tag{15}$$

then (12) lead to the equation of continuity,

$$\partial\mathfrak{J}_{\mu}/\partial x_{\mu}=0. \tag{16}$$

$\mathfrak{J}_{1,2,3}$ and $\mathfrak{J}_{4}$ are respectively the isotopic spin current density and isotopic spin density of the system. The equation of continuity guarantees that the total isotopic spin

$$\mathbf{T}=\int\mathfrak{J}_{4}d^{3}x$$

[8]Repeated indices are summed over, except where explicitly stated otherwise. Latin indices are summed from 1 to 3, Greek ones from 1 to 4.

is independent of time and independent of a Lorentz transformation. It is important to notice that $\mathfrak{J}_\mu$, like $\mathbf{b}_\mu$, does not transform exactly like vectors under isotopic space rotations. But the total isotopic spin,

$$\mathbf{T} = -\int \frac{\partial \mathbf{f}_{4i}}{\partial x_i} d^3x,$$

is the integral of the divergence of $\mathbf{f}_{4i}$, which transforms like a true vector under isotopic spin space rotations. Hence, under a general isotopic gauge transformation, if $S \to S_0$ on an infinitely large sphere, $\mathbf{T}$ would transform like an isotopic spin vector.

Equation (15) shows that the isotopic spin arises both from the spin-$\frac{1}{2}$ field ($\mathbf{J}_\mu$) and from the $\mathbf{b}_\mu$ field itself. Inasmuch as the isotopic spin is the source of the $\mathbf{b}$ field, this fact makes the field equations for the $\mathbf{b}$ field nonlinear, even in the absence of the spin-$\frac{1}{2}$ field. This is different from the case of the electromagnetic field, which is itself chargeless, and consequently satisfies linear equations in the absence of a charged field.

The Hamiltonian derived from (11) is easily demonstrated to be positive definite in the absence of the field of isotopic spin $\frac{1}{2}$. The demonstration is completely identical with the similar one in electrodynamics.

We must complete the set of equations of motion (12) and (13) by the supplementary condition,

$$\partial \mathbf{b}_\mu / \partial x_\mu = 0, \tag{17}$$

which serves to eliminate the scalar part of the field in $\mathbf{b}_\mu$. This clearly imposes a condition on the possible isotopic gauge transformations. That is, the infinitesimal isotopic gauge transformation $S = 1 - i\tau \cdot \delta\omega$ must satisfy the following condition:

$$2\mathbf{b}_\mu \times \frac{\partial}{\partial x_\mu} \delta\omega + \frac{1}{\epsilon} \frac{\partial^2}{\partial x_\mu^2} \delta\omega = 0. \tag{18}$$

This is the analog of the equation $\partial^2\alpha/\partial x_\mu^2 = 0$ that must be satisfied by the gauge transformation $A'_\mu = A_\mu + e^{-1}(\partial\alpha/\partial x_\mu)$ of the electromagnetic field.

Quantization

To quantize, it is not convenient to use the isotopic gauge invariant Lagrangian density (11). This is quite similar to the corresponding situation in electrodynamics and we adopt the customary procedure of using a Lagrangian density which is not obviously gauge invariant:

$$\begin{aligned}\mathcal{L}=&-\frac{1}{2}\frac{\partial \mathbf{b}_\mu}{\partial x_\nu}\cdot\frac{\partial \mathbf{b}_\mu}{\partial x_\nu}+2\epsilon(\mathbf{b}_\mu\times\mathbf{b}_\nu)\frac{\partial \mathbf{b}_\mu}{\partial x_\nu}\\&-\epsilon^2(\mathbf{b}_\mu\times\mathbf{b}_\nu)^2+\mathbf{J}_\mu\cdot\mathbf{b}_\mu-\bar{\psi}(\gamma_\mu\partial_\mu+m)\psi.\end{aligned}\tag{19}$$

The equations of motion that result from this Lagrangian density can be easily shown to imply that

$$\frac{\partial^2}{\partial x_\nu^2}\mathbf{a}+2\epsilon\mathbf{b}_\nu\times\frac{\partial}{\partial x_\nu}\mathbf{a}=0,$$

$$\mathbf{a}=\partial\mathbf{b}_\mu/\partial x_\mu.$$

Thus if, consistent with (17), we put on one space-like surface $\mathbf{a}=0$ together with $\partial\mathbf{a}/\partial t=0$, it follows that $\mathbf{a}=0$ at all times. Using this supplementary condition one can easily prove that the field equations resulting from the Lagrangian densities (19) and (11) are identical.

One can follow the canonical method of quantization with the Lagrangian density (19). Defining

$$\mathbf{\Pi}_\mu=-\partial\mathbf{b}_\mu/\partial x_4+2\epsilon(\mathbf{b}_\mu\times\mathbf{b}_4),$$

one obtains the equal-time commutation rule

$$\left[b_\mu^i(x),\Pi_\nu^j(x')\right]_{t-t'}=-\delta_{ij}\delta_{\mu\nu}\delta^3(x-x'),\tag{20}$$

where b_μ^i, $i=1,2,3$, are the three components of $\mathbf{b}_\mu$. The relativistic invariance of these commutation rules follows from the general proof for canonical methods of quantization given by Heisenberg and Pauli.[9]

The Hamiltonian derived from (19) is identical with the one from (11), in virtue of the supplementary condition. Its density is

$$\begin{aligned}H=&H_0+H_{\text{int}},\\H_0=&-\tfrac{1}{2}\mathbf{\Pi}_\mu\cdot\mathbf{\Pi}_\mu+\frac{1}{2}\frac{\partial\mathbf{b}_\mu}{\partial x_j}\cdot\frac{\partial\mathbf{b}_\mu}{\partial x_j}+\bar{\psi}(\gamma_j\partial_j+m)\psi,\\H_{\text{int}}=&2\epsilon(\mathbf{b}_i\times\mathbf{b}_4)\cdot\mathbf{\Pi}_i-2\epsilon(\mathbf{b}_\mu\times\mathbf{b}_j)\cdot(\partial\mathbf{b}_\mu/\partial x_j)\\&+\epsilon^2(\mathbf{b}_i\times\mathbf{b}_j)^2-\mathbf{J}_\mu\cdot\mathbf{b}_\mu.\end{aligned}\tag{21}$$

The quantized form of the supplementary condition is the same as in quantum electrodynamics.

[9] W. Heisenberg and W. Pauli, Z. Physik **56**, 1 (1929).

Properties of the b Quanta

The quanta of the **b** field clearly have spin unity and isotopic spin unity. We know their electric charge too because all the interactions that we proposed must satisfy the law of conservation of electric charge, which is exact. The two states of the nucleon, namely proton and neutron, differ by charge unity. Since they can transform into each other through the emission or absorption of a **b** quantum, the latter must have three charge states with charges $\pm e$ and 0. Any measurement of electric charges of course involves the electromagnetic field, which necessarily introduces a preferential direction in isotopic space at all space-time points. Choosing the isotopic gauge such that this preferential direction is along the z axis in isotopic space, one sees that for the nucleons

$$Q = \text{electric charge} = e(\tfrac{1}{2} + \epsilon^{-1} T^z),$$

and for the **b** quanta

$$Q = (e/\epsilon) T^z.$$

The interaction (7) then fixes the electric charge up to an additive constant for all fields with any isotopic spin:

$$Q = e(\epsilon^{-1} T^z + R). \tag{22}$$

The constants R for two charge conjugate fields must be equal but have opposite signs.[10]

We next come to the question of the mass of the **b** quantum, to which we do not have a satisfactory answer. One may argue that without a nucleon field the Lagrangian would contain no quantity of the dimension of a mass, and that therefore the mass of the **b** quantum in such a case is zero. This argument is however subject to the criticism that, like all field theories, the **b** field is beset with divergences, and dimensional arguments are not satisfactory.

One may of course try to apply to the **b** field the methods for handling infinities developed for quantum electrodynamics. Dyson's approach[11] is best suited for the present case. One first transforms into the interaction representation in which the state vector Ψ satisfies

$$i\partial\Psi/\partial t = H_{\text{int}}\Psi,$$

where H_{int} was defined in Eq. (21). The matrix elements of the scattering matrix are then formulated in terms of contributions from Feynman diagrams. These diagrams have three elementary types of vertices illustrated in Fig. 1, instead of only one type as in quantum electrodynamics. The "primitive divergences"

[10] See M. Gell-Mann, Phys. Rev. **92**, 833 (1953).
[11] F. J. Dyson, Phys. Rev. **75**, 486, 1736 (1949).

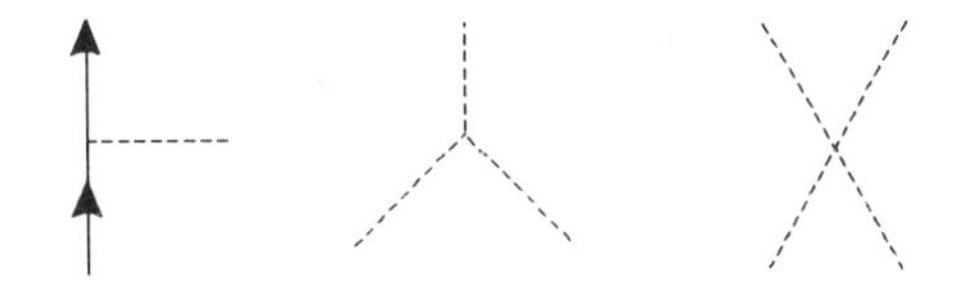

Figure 1. Elementary vertices for **b** fields and nucleon fields. Dotted lines refer to **b** field, solid lines with arrow refer to nucleon field.

are still finite in number and are listed in Fig. 2. Of these, the one labeled *a* is the one that effects the propagation function of the **b** quantum, and whose singularity determines the mass of the **b** quantum. In electrodynamics, by the requirement of electric charge conservation,[12] it is argued that the mass of the photon vanishes. Corresponding arguments in the **b** field case do not exist[13] even though the conservation of isotopic spin still holds. We have therefore not been able to conclude anything about the mass of the **b** quantum.

A conclusion about the mass of the **b** quantum is of course very important in deciding whether the proposal of the existence of the **b** field is consistent with experimental information. For example, it is inconsistent with present experiments to have their mass less than that of the pions, because among other

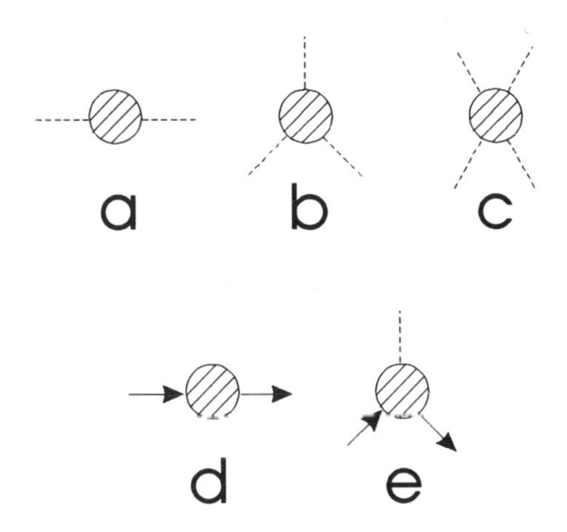

Figure 2. Primitive divergences.

[12] J. Schwinger, Phys. Rev. **76**, 790 (1949).

[13] In electrodynamics one can formally prove that $G_{\mu\nu}k_\nu = 0$, where $G_{\mu\nu}$ is defined by Schwinger's Eq. (A12). ($G_{\mu\nu}A_\nu$ is the current generated through virtual processes by the arbitrary external field A_ν.) No corresponding proof has been found for the present case. This is due to the fact that in electrodynamics the conservation of charge is a consequence of the equation of motion of the electron field alone, quite independently of the electromagnetic field itself. In the present case the **b** field carries an isotopic spin and destroys such general conservation laws.

reasons they would then be created abundantly at high energies and the charged ones should live long enough to be seen. If they have a mass greater than that of the pions, on the other hand, they would have a short lifetime (say, less than 10^{-20} sec) for decay into pions and photons and would so far have escaped detection.

9

Shaw's $SO(2)$ Approach, 1954–55

About the time that Yang was presenting his talk on the Yang-Mills theory at Princeton, the theory was being rediscovered quite independently by Ronald Shaw, who at that time was a postgraduate student of Abdus Salam at Cambridge University. It is interesting to record Shaw's contribution [7] because, although it came shortly after that of Yang and Mills, it was motivated in a completely different way. Through a preprint of Schwinger's, Shaw had become aware that electromagnetism could be formulated using the gauge group $SO(2)$, which is just the usual gauge group $U(1)$ quotiented by Z_2, and this had led him to a successful generalization from $SO(2)$ to $SU(2)$.

It is, perhaps, better to let Shaw speak for himself. In a letter to Kemmer, dated 26th May 1982, written in reply to a request by Kemmer for historical information on the subject, Shaw wrote:

> I am absolutely astonished how much gauge-fields have come to the fore in recent years. The idea seemed to me at the time as completely obvious: I had been reading (in 1953) some manuscript of Schwinger's in which he introduced the electromagnetic interaction in this way—he used real spinors and so had $SO(2)$, rather than $U(1)$, invariance and the generalization to $SU(2)$ invariance seemed to shout itself out!

In a later letter to Kemmer, dated August 1982, he added:

> What I am absolutely sure about is that my gauge-field work arose out of my fascination with invariance ideas in general prodded by a (rather rough) preprint of Schwinger's which I found lying around in 1953.

Later still, in a letter to Mills dated February 5th, 1985, he wrote:

> The idea arose in a flash, directly from reading some preprint of Schwinger's... I remember that the Schwinger notes were on rather poor quality paper, possibly cyclostyled in purple ink (but can one trust one's memory so far back). My memory does not quite match up with Schwinger's Phys. Rev. 91 (1953) 713 paper ... though he does there have
>
> $$\chi'(x) = \exp\{-i\lambda(x)\epsilon\}\,\chi(x), \quad \epsilon \quad a \text{ matrix}. \tag{168}$$
>
> I wonder if his notes were circulated but never published in the form in

> which I read them? ... At any rate it seemed to me an obvious idea to replace the $SO(2) \simeq U(1)$ of electromagnetism by $SU(2)$ (of Kemmer etc.) isospin and see what would happen. But I was disappointed that Nature (no suitable $m = 0$ particle) seemed to reject the idea.

Shaw's thesis consisted of two parts. Part I dealt with representations of the extended Lorentz group and their applications to particle physics, and Part II dealt with particle physics interactions. Part II contained three chapters, and it was in the last of these that Shaw constructed $SU(2)$ gauge theory. The construction was almost identical with that of Yang and Mills. He first noted that the usual isotopic (rigid $SU(2)$) invariance of the kinetic part $\bar{\psi}\gamma_\mu \partial^\mu \psi$ of the Lagrangian, where ψ is an isotopic 2-spinor, was lost if the isotopic spin transformations $\psi \to U\psi$ were made local. But he found that it could be restored by introducing a new vector-isovector field B^a_μ and changing the ordinary derivative ∂_μ to the covariant derivative $D_\mu = \partial_\mu + qB^a_\mu \tau_a$, where q is a coupling constant and the τ's are the generators of isospin transformation in the 2-dimensional representation, provided that the B fields had the transformation property

$$\vec{B}_\mu.\vec{\tau} \quad \to \quad U\left(\partial_\mu + q\vec{B}_\mu.\vec{\tau}\right)U^{-1}. \tag{169}$$

Like Yang and Mills, he was then faced by the problem of finding a kinetic term for the B fields. He first noted that the obvious choice, $\mathrm{tr}(\vec{f}_{\mu\nu} \cdot \vec{f}^{\mu\nu})$, where $\vec{f}_{\mu\nu} = \partial_\mu \vec{B}_\nu - \partial_\nu \vec{B}_\mu$, was not invariant. Then, without any further ado, he simply stated that

$$\mathrm{tr}\left(\vec{F}^{\mu\nu} \cdot \vec{F}_{\mu\nu}\right), \quad \text{where} \quad \vec{F}_{\mu\nu} = \partial_\mu \vec{B}_\nu - \partial_\nu \vec{B}_\mu - q\left(\vec{B}_\mu \wedge \vec{B}_\nu\right), \tag{170}$$

was invariant (see his equations (22)(23)). Having constructed the fermionic and B field Lagrangian in this way, he then derived the Yang-Mills field equations and conservation laws. Thus he followed essentially the same path as Yang-Mills but with a different motivation.

In trying to identify the B particles Shaw encountered the usual problems. He was convinced that the B field was massless, but, on the other hand, as he said,

> [Z]ero mass would appear to be ruled out since otherwise the neutron would have a rapid decay $N \to P + B^-$. In fact one would have expected the B-particles to have been observed unless their mass was quite large or unless the coupling constant q was very small.

His conviction that the B particles were massless does not quite agree with the statement on page 40 of his thesis. But this discrepancy was cleared up in the August 1982 letter to Kemmer in which he says:

> Incidentally, I see that p. 40 of my thesis (Part II) says that it is not clear that the mass of the B-field is zero. This may have been added after talking later on to Salam who was much more expert in such matters. Originally I did not doubt $m = 0$ and hence put aside my idea because it did not match up to physics.

The timing of Shaw's work is critical in deciding priorities. Luckily, the timing is clarified by Shaw himself in his August 1982 letter to Kemmer, where he says:

> I started research in the summer of 1952 and submitted my Ph.D. dissertation in September 1955. It consisted of two parts, and I am in no doubt at all that I did the work for part II first, completed sometime in 1954. This in itself consisted of several disjoint bits, including Ch.III on $SU(2)$ gauge-fields. I remember feeling inadequate ... and so I searched around and in late 1954 and in 1955 produced Part I of my thesis.
> According to the footnote on p. 37 of Part II, my $SU(2)$ gauge-field work was completed by January 1954. Sometime later (still early in 1954?) I showed it to Salam but I do not think that he appreciated it at the time. I do not at all blame him for this—I probably told him about it in a very dismissive way, since the relevant particles, surely $m = 0$ I thought, did not exist. Much later (end of 1954?) he told me that Yang and Mills had had the same idea and told me to write mine up (which I did not).

Thus there is no doubt about Yang and Mills's priority. They had completed the theory in the summer of 1953 and had it published in a regular journal by April 1954. Shaw had completed his theory in January 1954 and had it published as a thesis in September 1955. But it was a close race.

Finally it should be recorded that Shaw is

> very anxious to remove any possible suggestion that Salam did not publicize the contribution sufficiently. On the contrary, he (Salam) frequently announced the contribution at conferences—e.g. Istanbul 1962 Summer School (on Group Theory in Physics) and the Schrödinger Centenary Conference at Imperial College 1987. He also often referred to the 'Yang-Mills-Shaw theories' e.g. in his Nobel Prize talk.

Invariance under General Isotopic Spin Transformations

by Ronald Shaw in Cambridge

Excerpted from "The Problem of Particle Types and Other Contributions to the Theory of Elementary Particles" (Ph.D. diss., Cambridge University, 1955), pt. II, ch. III.

1. Introduction

Many important general results follow from postulating that the Lagrangian of the system of fields should be invariant under certain classes of transformations. For example, if the Lagrangian is invariant under translations and rotations in space-time, then it is well-known that a momentum 4-vector P_α and an angular momentum tensor $M_{\alpha\beta}$ exist and are conserved. These two invariances seem to exhaust the useful possibilities for transformations in space-time. However one can also postulate the invariance of the Lagrangian under rotations in a Euclidean space E_n of dimension n which has no relation to space-time, and useful results still follow.

The invariance of the Lagrangian under special gauge transformations can be looked upon from this point of view. The space involved is then E_2, and invariance under rotations in this space results in the existence of a divergenceless current vector s_α and of a charge Q which is conserved. Neutral fields are scalars in E_2, while fields with charge one are vectors in E_2. For instance, taking the case of two real fermion fields ψ_1 and ψ_2 with free Lagrangian*

$$L_o = \tfrac{1}{2}i(\psi_i B\gamma_\alpha \partial^\alpha \psi_i - \partial^\alpha \psi_i B\gamma_\alpha \psi_i) + im\psi_i B\psi_i \tag{1}$$

it is seen that L_o is invariant under the infinitesimal rotation

$$\begin{aligned} \psi_1' &= \psi_1 - c\psi_2 \\ \psi_2' &= \psi_2 + c\psi_1 \end{aligned} \tag{2}$$

In equation (1) B is the matrix defined by $B\gamma_\alpha B^{-1} = -\gamma_\alpha{}^T$; it then follows that $\psi B\psi$ is a scalar, $\psi B\gamma^\alpha\psi$ a vector, etc. The current density and total charge

*Lorentz suffices are represented by Greek letters, and suffices in E_n by Latin letters. The summation convention is employed with respect to both kinds of suffices.

are in this case

$$s^\alpha = ie(\psi_1 B\gamma^\alpha \psi_2 - \psi_2 B\gamma^\alpha \psi_1) \tag{3}$$

$$Q = \int s^\alpha \, d\sigma_\alpha \tag{4}$$

In the special gauge transformation (2), c was a constant independent of the space-time argument of ψ. If now the Lagrangian is required to be invariant under general gauge transformations (i.e. those with c a function of position) then it is found necessary to introduce the electromagnetic field. This method was adopted by J.Schwinger (Phys. Rev. **91** (1953) 713). For under general gauge transformations, L_0 is no longer invariant but becomes

$$L_0' = L_0 - i\partial_\alpha c(\psi_1 B\gamma^\alpha \psi_2 - \psi_2 B\gamma^\alpha \psi_1) \tag{5}$$

However the last term of (5) can be compensated by taking an interaction Lagrangian

$$L_1 = -A_\alpha s^\alpha \tag{6}$$

and requiring that when ψ undergoes the transformation (2), A_α should undergo the transformation

$$eA_\alpha' = eA_\alpha + \partial_\alpha c \tag{7}$$

Finally, defining as usual the field strengths

$$f_{\alpha\beta} = \partial_\alpha A_\beta - \partial_\beta A_\alpha \tag{8}$$

and a free Lagrangian L_2 for the electromagnetic field

$$L_2 = \tfrac{1}{4} f_{\alpha\beta} f^{\alpha\beta} \tag{9}$$

the total Lagrangian

$$L = L_0 + L_1 + L_2 \tag{10}$$

is seen to be invariant under general gauge transformations.

The purpose of the above treatment of the transition from special gauge invariance to general gauge invariance was to present the introduction of the electromagnetic field in a form suitable for generalisation to spaces E_n with $n > 2$. For in the usual symmetrical meson theory (N.Kemmer, Proc. Camb. Phil. Soc. **34** (1938) 354), invariance under rotations in a space E_3, the so-called isotopic spin space, is assumed. An infinitesimal rotation in this space involves three real parameters c_i instead of the one parameter c of the transformation (2). In the usual theory the c_i do not depend on position, or in other words, only special isotopic spin transformations are contemplated. However, in analogy with the case of E_2, it is natural to postulate invariance of the Lagrangian under general isotopic spin transformations, the c_i now being functions of position. The consequences* of this postulate are worked out in Section 2. In Section 3, the Salam-Polkinghorne scheme, in which all strong interactions are assumed to be invariant under rotations in a space E_4, is generalised in a similar manner.

2. Invariance in the space E_3

The fermion case is taken once more, with free field Lagrangian

$$L_0 = \bar{\psi}(\gamma_\alpha \partial^\alpha + m)\psi \tag{11}$$

where ψ is 4-component spinor in space-time and a 2-component spinor in E_3. Note that this is in contrast to the case of E_2, when ψ was taken to be a vector in E_2. In (11) the two isotopic spin components of ψ refer of course to the proton and neutron. Under a rotation

$$\phi_i' = \phi_i + c_{ij}\phi_j \tag{12}$$

in isotopic spin space, ψ undergoes the transformation

$$\psi' = (1 + \tfrac{1}{2} i c_i \tau_i)\psi \tag{13}$$

where $c_i = \frac{1}{2}\varepsilon_{ijk} c_{jk}$ and where the τ_i are the usual isotopic spin matrices. In analogy with (3) and (4) the isotopic spin current density $\vec{s}_\alpha$ and the total isotopic spin $\vec{T}$ can be defined

$$\vec{s}_\alpha = +\tfrac{1}{2} i q \bar{\psi} \gamma_\alpha \vec{\tau} \psi \tag{14}$$

and

$$\vec{T} = \int \vec{s}_\alpha \, d\sigma^\alpha; \tag{15}$$

$\vec{s}_\alpha$ is seen to be divergenceless and hence $\vec{T}$ is conserved. In (14) q is the unit of "isotopic charge" and corresponds to e in (3).

Now under a general transformation (13) with $\vec{c}$ a function of position, L_0 is no longer invariant but becomes

$$L_0' = L_0 + \tfrac{1}{2} i \partial^\alpha \vec{c} \cdot \bar{\psi} \gamma_\alpha \vec{\tau} \psi. \tag{16}$$

*The work described in this chapter was completed, except for its extension in Section 3, in January 1954, but was not published. In October 1954, Yang and Mills adopted independently the same postulate and derived similar consequences.

The addtional term can be compensated by introducing an interaction

$$L_1 = -\vec{B}^\alpha \cdot \vec{s}^\alpha \tag{17}$$

and by requiring that when ψ transforms under (13) the field $\vec{B}^\alpha$, which is the generalisation to E_3 of the electromagnetic field, transforms under

$$q\vec{B}'^\alpha = q(\vec{B}^\alpha - \vec{c} \times \vec{B}^\alpha) + \partial^\alpha \vec{c}. \tag{18}$$

In order to obtain a theory invariant under isotopic spin transformations it only remains to construct an invariant free field Lagrangian for the $\vec{B}$-field. It is natural, in analogy with the electromagnetic case, to define

$$\vec{f}^{\alpha\beta} = \partial^\alpha \vec{B}^\beta - \partial^\beta \vec{B}^\alpha \tag{19}$$

and to take the Lagrangian to be

$$L_2 = -\tfrac{1}{4}\vec{f}^{\alpha\beta} \cdot \vec{f}_{\alpha\beta} \tag{20}$$

However (20) is not invariant under the transformation (18). For $\vec{f}^{\alpha\beta}$ transforms as follows

$$\vec{f}'^{\alpha\beta} = \vec{f}^{\alpha\beta} - \vec{c} \times f^{\alpha\beta} - 2(\partial^\alpha \vec{c} \times \vec{B}^\beta - \partial^\beta \vec{c} \times \vec{B}^\alpha) \tag{21}$$

The second order derivatives of $\vec{c}$ have cancelled as in the electromagnetic case, but some first order derivatives still remain and spoil the invariance of (20). A properly invariant Lagrangian can be obtained by defining

$$\vec{F}^{\alpha\beta} = \vec{f}^{\alpha\beta} - q(\vec{B}^\alpha \times \vec{B}^\beta - \vec{B}^\beta \times \vec{B}^\alpha) \tag{22}$$

and by taking L_2 to be

$$L_2 = -\tfrac{1}{4}\vec{F}^{\alpha\beta} \cdot \vec{F}_{\alpha\beta} \tag{23}$$

instead of (20). L_2 is now invariant under (18), as $\vec{F}^{\alpha\beta}$ transforms solely as a vector in E_3.

The field equations which follow from the Lagrangian

$$L = L_0 + L_1 + L_2 \tag{24}$$

are found to be

$$(\gamma_\alpha \partial^\alpha + m)\psi = -\tfrac{1}{2} i q \vec{B}^\alpha \cdot \vec{\tau} \gamma_\alpha \psi \tag{25}$$

and

$$-\partial_\beta \vec{F}^{\alpha\beta} + 2q\vec{B}_\beta \times \vec{F}^{\alpha\beta} = \vec{s}^\alpha \tag{26}$$

The divergence of $\vec{s}^\alpha$ no longer vanishes, but a divergenceless vector $\vec{j}^\alpha$, say, follows immediately from (26)

$$\vec{j}^\alpha = \vec{s}^\alpha - 2q\vec{B}_\beta \times \vec{F}^{\alpha\beta} \tag{27}$$

and instead of (15), the total isotopic spin is now

$$\vec{T} = \int \vec{j}^\alpha \, d\sigma_\alpha. \tag{28}$$

$\vec{j}^\alpha$ replaces $\vec{s}^\alpha$ as the isotopic spin current density of the system, and is composed of the isotopic spin $\vec{s}^\alpha$ of the nucleons and the isotopic spin $-2q\vec{B}_\beta \times \vec{F}^{\alpha\beta}$ of the B-field. The fact that the B-field itself carries isotopic spin accounts for the greater complexity of equation (26) compared with the corresponding equation $-\partial_\beta f^{\alpha\beta} = s^\alpha$ in the electromagnetic case, and it is not clear from (26) whether the mass of the B-field is zero or not. This means that it is difficult to determine whether or not the existence of B-particles is consistent with experiment. It is clear that these particles, if they exist, will have spin 1 and isotopic spin 1, and so exist in three states with charge $\pm e$ and 0, like the π-meson. Zero mass would appear to be ruled out since otherwise the neutron would have a rapid decay

$$N \to P + B^- \tag{29}$$

In fact one would have expected the B-particles to have been observed unless their mass was quite large or unless the coupling constant q was very small.

Finally it should be pointed out that the invariance postulated in this section does not hold when the electromagnetic field is taken into account. The status of invariances like this, which apply only to part of the Lagrangian, is obscure and is to be contrasted with invariance under gauge transformations, where the ensuing conservation law, that of electric charge, seems never to be violated. The existence of this contrast was the motive of the above work, but the conclusion is that gauge invariance and isotopic spin invariance do not appear to differ essentially as far as generalisation to space-dependent transformations is concerned.

3. Invariance in the space E_4

In this section the 4-dimensional theory of A. Salam and J. Polkinghorne (Nuovo Cim. **2** (1955) 685) is generalised, in a way similar to the 3-dimensional isotopic spin case of Section 2. The resulting equations are slightly more cumbersome as vector notation cannot be employed.

The fermion free field Lagrangian is now

$$L_0 = \bar{\Psi}_i(\gamma_\alpha \partial^\alpha + m)\Psi_i \tag{30}$$

where Ψ_i is a four-vector in E_4 and describes both nucleons and cascade particles. The connection between Ψ_i and these fields can be taken as

$$\begin{aligned} 2^{\frac{1}{2}}\psi_P &= \Psi_0 + i\Psi_3, \quad 2^{\frac{1}{2}}\psi_N = \Psi_1 - i\Psi_2 \\ 2^{\frac{1}{2}}\chi_{\Xi^-} &= \Psi_0 - i\Psi_3, \quad 2^{\frac{1}{2}}\chi_{\Xi^o} = \Psi_1 + i\Psi_2 \end{aligned} \tag{31}$$

In this case the interaction with the electromagnetic field is

$$L' = -e(\bar{\Psi}_0\gamma_\alpha\Psi_3 - \bar{\Psi}_3\gamma_\alpha\Psi_0)A^\alpha \tag{32}$$

In this scheme the π-meson field transforms under the representation $R(1,0)$ of the rotation group in E_4, and so is a self dual antisymmetric tensor. Its interaction with nucleons and cascade particles is taken to be

$$L'' = ig\bar{\Psi}_i\gamma_5\Psi_j\,\phi_{ij} \tag{33}$$

This can be written in the form

$$L'' = ig\bar{\Psi}\gamma_5\vec{T}\Psi\cdot\vec{\phi} \tag{34}$$

where $\vec{T}$ is the self-dual part of the infinitesimal generators M_{ij} of the vector representation in E_4; that is $T_1 = \frac{1}{2}(M_{01} + M_{23})$ etc. Since the T_i satisfy the commutation relations for generators of the spin $\frac{1}{2}$ representation in E_3, we can take

$$-2i\,\vec{T} = \left\| \begin{matrix} \vec{\tau} & 0 \\ 0 & \vec{\tau} \end{matrix} \right\| \tag{35}$$

in a suitable representation and so (34) can be written in the form

$$L'' = g\,\vec{\phi}\cdot(\bar{\psi}\gamma_5\,\vec{\tau}\psi + \bar{\chi}\gamma_5\,\vec{\tau}\chi) \tag{36}$$

where ψ is the nucleon wave function with components ψ_P and ψ_N, and where χ is the cascade particle wave function with components χ_{Ξ^o} and χ_{Ξ^-}. Thus the above scheme gives the usual symmetrical interaction between π-mesons and nucleons.

Under infinitesimal rotations in E_4, Ψ transforms

$$\Psi_i' = \Psi_i + c_{ij}\Psi_j. \tag{37}$$

If invariance is postulated when the c_{ij} are functions of position, then it is necessary to add to L_0 an interaction Lagrangian

$$L_1 = -\tfrac{1}{2} q B^{\alpha}_{\ ij}(\bar{\Psi}_i \gamma_\alpha \Psi_j - \bar{\Psi}_j \gamma_\alpha \Psi_i) \tag{38}$$

where $B^{\alpha}_{\ ij}$ transforms under (37) as follows

$$q B'^{\alpha}_{\ ij} = q(B^{\alpha}_{\ ij} + c_{ij} B^{\alpha}_{\ kj} - c_{jk} B^{\alpha}_{\ ki}) + \partial^\alpha c_{ij} \tag{39}$$

The invariant free Lagrangian for the B-field is

$$L_2 = -\tfrac{1}{4} F^{\alpha\beta}_{ij}\, F^{ij}_{\alpha\beta}, \tag{40}$$

where the “field strengths” $F^{\alpha\beta}_{ij}$ are defined by

$$F^{\alpha\beta}_{ij} = \partial^\alpha B^{\beta}_{\ ij} - \partial^\beta B^{\alpha}_{\ ij} - q(B^{\alpha}_{\ ik} B^{\beta}_{\ kj} - B^{\alpha}_{\ jk} B^{\beta}_{\ ki}) \tag{41}$$

The field equations are easily derived and are

$$(\gamma_\alpha \partial^\alpha + m)\Psi_i = q B^{\alpha}_{\ ij} \gamma_\alpha \Psi_j \tag{42}$$

and

$$-\partial_\beta F^{\alpha\beta}_{ij} + q(B_{ik,\beta} F^{\alpha\beta}_{kj} - B_{jk,\beta} F^{\alpha\beta}_{ki}) = s^{\alpha}_{ij} \tag{43}$$

where the fermion isotopic spin current density $s^{\alpha}_{\ ij}$ is given by

$$s^{\alpha}_{\ ij} = q(\bar{\Psi}_i \gamma^\alpha \Psi_j - \bar{\Psi}_j \gamma^\alpha \Psi_i) \tag{44}$$

The total isotopic spin is

$$T_{ij} = \int j^{\alpha}_{ij}\, d\sigma_\alpha \tag{45}$$

where

$$j^{\alpha}_{\ ij} = s^{\alpha}_{\ ij} - q(B_{ik,\beta}\, F^{\alpha\beta}_{kj} - B_{jk,\beta}\, F^{\alpha\beta}_{ki}) \tag{46}$$

The divergence of $j^{\alpha}_{\ ij}$ vanishes, and so T_{ij} is conserved.

An interesting result of the above theory is obtained on comparing the interactions (32) and (38). It is seen that the electromagnetic interaction L' can be considered as part of the B-field interaction L_1 as long as $q = e$ and $A^\alpha = B^\alpha_{03}$. The total charge is then $Q = T_{03}$. In fact if the B-field were non-quantised, then one could start from the Lagrangian $L = L_0 + L_1 + L_2$, invariant under general transformations (37) and (39), and apply such a transformation at each point in space-time, in such a way as to make all components of B^α_{ij} vanish except B^α_{03}. With the above identification it is seen that the field equations (42) and (43) would then be exactly those for nucleons and cascade particles in interaction with the electromagnetic field,* so that while apparently starting with a more general theory, the usual theory is still obtained. Of course the above reduction of B_{ij} to B_{03} could not be carried out in a quantised theory, and so one would expect the existence of six B-particles of spin 1. Since B_{ij} is an antisymmetric tensor, it transforms under the reducible representation $R(1, 0) + R(0, 1)$, and so it is natural to suppose that the above six particles should exist in two sets of three with wave functions $\vec{C}^\alpha$ and $\vec{D}^\alpha$, say, where $C^\alpha_1 = \frac{1}{2}(B^\alpha_{01} + B^\alpha_{23})$, $D^\alpha_1 = \frac{1}{2}(B^\alpha_{01} - B^\alpha_{23})$ etc. The interaction (38) can be written in terms of the nucleon, cascade, C and D fields. The interaction with the C-field is

$$q(\bar{\psi}\gamma_\alpha\vec{\tau}\psi + \bar{\chi}\gamma_\alpha\vec{\tau}\chi)\cdot\vec{\sigma}^\alpha$$

and with the D-field is

$$q(\bar{\psi}\gamma_\alpha\chi + \bar{\chi}\gamma_\alpha\psi)D^\alpha_1 + iq(-\bar{\psi}\gamma_\alpha\chi + \bar{\chi}\gamma_\alpha\psi)D^\alpha_2 + q(\bar{\psi}\gamma_\alpha\psi - \bar{\chi}\gamma_\alpha\chi)D^\alpha_3$$

The C-field has the π-meson type interaction in the Salam scheme while the D-field has an interaction of the τ-meson type (except that a vector interaction rather than a pseudoscalar one has to be taken). However, if the B-field does in fact exist, it is not certain that it will be observed in its C and D forms separately, for one would hope that the combination $C_3 + D_3$ would be the electromagnetic field.

*Except that the mass difference between nucleons and cascade particles is not taken into account. This large mass difference seems to be the only real objection to the Salam-Polkinghorne theory.

10

Utiyama's General Approach, 1954–55

Meanwhile, a completely independent approach to non-abelian gauge theories was being adopted by Utiyama in Japan. Utiyama's motivation was to find a structure that would be common to gravitation and the interactions of the electromagnetic type. The structure he found was the *connection.* Utiyama does not quote Weyl, and it is not clear whether he had read the 1929 paper, but his treatment of the problem closely followed the classical pattern of Weyl's 1929 paper. He effectively extended Weyl's gauge principle to general Lie groups. As he also included gravity, it is fair to say that Utiyama's approach was the broadest and most comprehensive. But from the point of view of priority, his contribution appeared later than that of Yang and Mills and slightly later than Shaw's, though again it was a close finish. Utiyama actually found his results early in 1954 and presented them at a workshop held at the Yukawa Institute, Kyoto, in the early summer of that year but did not send them for publication until late 1955. How this came about is an interesting story in itself and has been described by Utiyama in one of his books [8]. As far as I know, this book is available only in Japanese and Russian, and I am grateful to Izumi Tutsui and Mikhail Saveliev for extracting the relevant statements and putting them in chronological order:

> Chapter 10, entitled Memoir of Regret:
>
> January 1954: At the end of January I (R. Utiyama) received a letter from the Institute of Advanced Study (IAS), Princeton, inviting me to visit the IAS from September 1954. I decided to prepare some completed work prior to going so that I could publish papers even if I could not work successfully there. In fact, I had already the idea of the *general gauge-theory* in mind, including the basic structure of the theory, before I received the letter from the IAS. So what I should do was simply to spell it out in the form of a paper. I completed the work at the end of March 1954.
>
> May or June 1954: I gave a talk on the work at a small workshop held at the then Research Institute for Fundamental Physics, Kyoto University [now Yukawa Institute for Theoretical Physics]. I was disappointed by the response. [It appears that the main objections to Utiyama's theory were (i) that it did not seem to be in agreement with the Yukawa theory, and (ii) that

it was contrary to the historical precedents in which the gauge invariance followed from the physical law, not vice versa. But it seems that his talk provoked an active discussion, not all of which was negative.]

September 1954: I arrived at Princeton. I met Professor Minoru Kobayashi (then professor at Kyoto University) who was my supervisor at Osaka University and had been staying at the IAS from one year earlier as a visiting fellow. When I explained to him my idea for a general gauge theory, he told me that recently Professor Yang had announced a theory that was quite similar to mine. Some days later, I found a preprint by Yang in my mailbox. I took a look at it and found many formulae already familiar to me—I immediately realized that he had found the same theory as I had developed. I was too deeply shocked to examine Yang's paper closely and to compare it carefully with my work.

March 1955: As my appointment was prolonged for one more year, I felt that I should work on a subject which was more important than those I had engaged in so far at the IAS. So I decided to return to the general gauge theory and took a closer look at Yang's paper, which had been published in 1954. At this moment I realized for the first time that there was a significant difference between Yang's theory and mine. The difference was that Yang had merely found an example of non-abelian gauge-theory whereas I had developed a general idea of gauge-theory which would contain gravity as well as electromagnetic theory. Thus I decided to publish my work by translating it into English, and adding an extra section where Yang's theory is discussed as an example of my general Theory. I did so partly because his paper had already been published and partly because I thought that it would not be fair if I did not refer to his paper after I had read it, even if I had developed the idea of the general gauge theory before knowing of his work. My paper was accepted for publication in Physical Review in July 1955 and published early in 1956.

Remark: I admit that it was Yang and Mills who first found a new type of (non-abelian) gauge-theory (though not gravity). However, it is very disappointing to me to see that non-abelian gauge-theory is called by the name 'Yang-Mills theory' only, without the slightest mention of me who developed the general framework of the gauge theory. I very much regret not having submitted my paper to a Japanese journal in March 1954 when I had completed the work.

Utiyama is, of course, absolutely correct in emphasizing the generality of his construction. Because of his interest in gravity, he formulated the theory not only for simple compact Lie groups but for general Lie groups. None of the other authors mentioned, Klein, Pauli, Yang-Mills or Shaw, even considered going beyond the groups $U(1)$ and $SU(2)$. Indeed, the higher-rank semi-simple groups were almost unknown to particle physicists at the time (recall that flavor $SU(3)$ was discovered only seven years later), and this makes Utiyama's achievement all the more remarkable. He was also the only one to suggest that the broad analogy between gravitation and electromagnetism that had so fascinated Weyl, Einstein, Schrödinger and others could be extended to include all the interactions, or conversely, that the Yang-Mills type theory might be extended to encompass gravitation. As he says in another book [9]:

> If I am allowed to refer to my work on the general gauge-theory, I should like to stress that it is my paper which first showed clearly that the theory of gravitation could fall into the framework of the gauge theory, which would be a first step toward the grand unified theory from the modern viewpoint. Even more importantly, my paper pointed out that fields carrying a fundamental force—either gravity or electromagnetism—must in fact be those termed *connections* in mathematics, which are now called *gauge-potentials*. That the concept of connections is indispensable in establishing a theory of interactions was the basic assertion that I wanted to make in my paper.

Utiyama's paper, which was published in *Physical Review* in 1956, is reproduced here and speaks for itself. Nevertheless it might be worthwhile to draw attention to some important points.

First, in the introduction, he asked the now familiar question as to what happens when a Lagrangian which is invariant with respect to a rigid Lie group G is required to be invariant with respect to the corresponding space-time dependent group $G(x)$. In particular, he asked what kind of gauge field must be introduced to maintain the symmetry, how it should transform with respect to $G(x)$ and how the new Lagrangian should compare with the original one.

In section 1, he answered all these questions by showing (after some detailed computation) that the gauge field must be a space-time vector field $A_\mu(x)$ taking its values in the adjoint representation of the Lie group and that it had to transform so that the derivative,

$$D_\mu = \partial_\mu + A_\mu(x), \tag{171}$$

is covariant. Thus the original Lagrangian $L_m(Q(x), \partial_\mu Q(x))$ had to be changed to $L_m(Q(x), R(D_\mu)Q(x))$, where the $Q(x)$ are matter fields and $R(D_\mu)$ is the representative of D_μ in the representation of G to which the Q belong. Finally he showed that if the full Lagrangian were assumed to be of the form $L_m + L_o(A_\mu)$, where L_m is the Lagrangian for the matter fields, then the pure-gauge part $L_o(A_\mu)$ could be a function only of the field strengths

$F_{\mu\nu} = [D_\mu, D_\nu]$. In brief, he constructed non-abelian gauge theory along the same lines as Weyl had constructed abelian gauge theory in 1929.

In the following sections, 2, 3 and 4, Utiyama applied the theory to derive electromagnetism, Yang-Mills theory and gravitational theory, respectively. In the case of gravitation, he rederived all the old results, but starting from the gauge principle as it is used in the other interactions.

Utiyama has sometimes been criticized on the grounds that he emphasized the similarities between gravitation and other gauge theories without stressing the differences. However, although he did not stress them, the differences are exhibited by his results.

The most obvious differences have been already mentioned at the end of Chapter 5. These are (i) that for gravitation the connection is not fundamental but is derived from a metric, (ii) that there is no equivalence principle for the other interactions, and (iii) that the gravitational Lagrangian is linear rather than quadratic in the field strengths. But one of the merits of Utiyama's paper is that the comparison of gravitation with non-abelian gauge theory instead of electromagnetism exhibits the differences in a quantitative manner. Thus, it highlights the fact that, although the non-gravitational interactions have no universal equivalence principle, they have a limited equivalence principle in the sense that all fields belonging to the same irreducible representation of the gauge group couple to the gauge field in the same way. From this point of view, gravitation is distinguished only by the fact that all matter belongs to a single representation (carried by the energy-momentum tensor). The comparison also shows that, although the non-abelian Lagrangian differs from the gravitational one in being quadratic rather than linear in the field strengths, it resembles it qualitatively in that it produces a self-interaction of the gauge field, but differs from it quantitatively, i.e., in the details of the interaction.

Utiyama's remarks on the application of the theory are also worth recording. Calling interactions induced by the above gauge principle first-class interactions and all others second-class, he begins: "[T]he electromagnetic, gravitational and Yang-Mills interactions belong to the first-class and the meson-nucleon interaction to the second-class, at least at the present stage." But later, he continues:

> The above stated classification has only a tentative meaning. Some of the interactions of the second-class might be translated to the first class if we could find a transformation group by means of which we could derive that interaction following the general scheme of section 1. For example, if the interaction between mesons and nucleons could be re-interpreted in a fashion analogous to those of the first-class, then one might presumably be able to get a wider viewpoint for interpreting the interactions between the new unstable particles and the nucleons.

In view of these remarks, it is not surprising to hear that Utiyama was very pleased by the advent of color $SU(3)$ and quark theory since they did just what he had hoped—they permitted (indeed, forced) a reinterpretation of the meson-nucleon interactions as a gauge interaction and provided a broader view for interpreting the interactions between nucleons and unstable resonances

Invariant Theoretical Interpretation of Interaction

Ryoyu Utiyama in Princeton, N.J.

Phys. Rev. 101 (1956) 1597

Some systems of fields have been considered which are invariant under a certain group of transformations depending on n parameters. A general rule is obtained for introducing a new field in a definite way with a definite type of interaction with the original fields by postulating the invariance of these systems under a wider group derived by replacing the parameters of the original group with a set of arbitrary functions. The transformation character of this new field under the wider group is determined from the invariance postulate. The possible types of the equations of the new fields can be also derived, giving rise to a certain conservation law owing to the invariance. As examples, the electromagnetic, the gravitational and the Yang-Mills fields are reconsidered following this line of approach.

Introduction

The form of the interactions between some well known fields can be determined by postulating invariance under a certain group of transformations. For example, let us consider the electromagnetic interaction of a charged field $Q(x)$, $Q^*(x)$. The electromagnetic interaction appears in the Lagrangian through the expressions

$$\frac{\partial Q}{\partial x^\mu} - ieA_\mu Q \quad \text{or} \quad \frac{\partial Q^*}{\partial x^\mu} + ieA_\mu Q^*. \tag{1}$$

The gauge invariance of this system is easily verified in virtue of the combinations of Q, Q^*, and A_μ in (1), if this system is invariant under the phase transformation

$$Q \to e^{i\alpha} Q, \quad Q^* \to Q^* e^{-i\alpha}, \quad \alpha = \text{const.} \tag{2}$$

Reversing the argument, the combination (1) can be uniquely introduced by the following line of reasoning. In the first place, let us suppose that the Lagrangian $L(Q, Q_\mu)$ is invariant under the constant phase transformation (2). Let us replace this phase transformation with the wider one (gauge transformation) having the phase factor $\alpha(x)$ instead of the constant α. In order to make the Lagrangian still invariant under this wider transformation it is necessary to introduce the electromagnetic field through the combination (1). This combination and the transformation character of A_μ under the gauge transformation can be uniquely determined from the gauge invariance postulate of the Lagrangian $L(Q, Q_{,\mu}, A_\mu)$.

This approach was taken by Yang and Mills[1] to introduce their new field $\mathbf{B}_\mu$ which interacts with fields having nonvanishing isotopic spins. The gravitational interaction also can be introduced in this fashion.

It may be worthwhile to investigate this approach for a more general case, for if there is a system of fields $Q^A(x)$ which is invariant under some transformation group depending on parameters $\epsilon_1, \epsilon_2, \cdots \epsilon_n$, then according to the aforementioned viewpoint we may have the possibility of introducing a new field, say $A(x)$, in a definite way. In addition, the transformation character of this new field and the interaction form with the Q's can be determined uniquely.

Let us tentatively call a family of the interactions derived in this way "the interactions of the first class," while other types of interactions are denoted as "the interactions of the second class." The electromagnetic, gravitational and $\mathbf{B}_\mu$-field interactions belong to the first class and the meson-nucleon interaction to the second class, at least at the present stage.

The main purpose of the present paper is to investigate the following problem. Let us consider a system of fields $Q^A(x)$, which is invariant under some transformation group G depending on parameters $\epsilon_1, \epsilon_2, \cdots \epsilon_n$. Suppose that the aforementioned parameter-group G is replaced by a wider group G', derived by replacing the parameters ϵ's by a set of arbitrary functions $\epsilon(x)$'s, and that the system considered is invariant under this wider group G'. Then, can we answer the following questions by using only the postulate of invariance stated above? (1) What kind of field, $A(x)$, is introduced on account of the invariance? (2) How is this new field A transformed under G'? (3) What form does the interaction between the field A and the original field Q take? (4) How can we determine the new Lagrangian $L'(Q, A)$ from the original one $L(Q)$? (5) What type of field equations for A are allowable?

The solution of these problems will be stated in Sec. 1. In Secs. 2, 3, and 4 the well-known examples of the interactions of the first class will be reconsidered following the line of reasoning of Sec. 1. We shall find an analogy between the transformation characters of the electromagnetic field A_μ, the Yang-Mills field $\mathbf{B}_\mu$, and Christoffel's affinity $\Gamma_{\mu\nu}{}^\lambda$ in the theory of the general relativity. Furthermore we shall understand the reason why in the Yang-Mills field strength the quadratic term, $\mathbf{B}_\mu \times \mathbf{B}_\nu$, appears which is quite similar to that occurring in the Riemann-Christoffel tensor $R_{\mu\nu\rho}{}^\lambda$, namely, to the term $\Gamma\Gamma - \Gamma\Gamma$ in R.

In the usual textbooks of general relativity the covariant derivative of any tensor is introduced by using the concept of parallel displacement. On the

[1]C. N. Yang and R. L. Mills, Phys. Rev. **96**, 191 (1954).

other hand, we shall see in Sec. 4 that the covariant derivative of any tensor or spinor can be derived from the postulate of invariance under the "generalized Lorentz transformations" derived by replacing the six parameters of the usual Lorentz group with a set of six arbitrary functions of x. In deriving such covariant derivatives it is unnecessary to use explicitly the notion of parallel displacement.

Now the above stated classification of the interactions has only a tentative meaning. Some of the interactions of the second class might be translated to the first class if we could find a transformation group by means of which we can derive that interaction following the general scheme in Sec. 1. For example, if the interaction between mesons and the nucleons could be reinterpreted in a fashion analogous to those of the first class, then one might presumably be able to get a wider viewpoint for interpreting the interactions between the new unstable particles and the nucleons.

1. General Theory

Let us consider a set of fields $Q^A(x)$, $(A = 1, 2, \ldots N)$, with the Lagrangian density

$$L(Q^A, Q^A{}_{,\mu}), \quad Q^A{}_{,\mu} = \partial Q^A / \partial x^\mu.$$

Now let us postulate that the action integral referred to some arbitrary four-dimensional domain Ω,

$$I = \int_\Omega L d^4 x,$$

is invariant under the following infinitesimal transformation:

$$\begin{aligned} Q^A &\to Q^A + \delta Q^A, \\ \partial Q^A &= T_{(a),}{}^A{}_B \epsilon^a Q_B, \\ \epsilon^a &= \text{infinitesimal parameter } (a = 1, 2, \ldots n), \\ T_{(a),}{}^A{}_B &= \text{constant coefficient.} \end{aligned} \tag{1.1}$$

In addition, the transformation (1.1) is assumed to be a Lie group G depending on the n parameters ϵ^a.

Thus there must be a set of constants $f_b{}^a{}_c$ called the "structure constants," which are defined by

$$[T_{(a)}, T_{(b)}]^A{}_B = T_{(a),}{}^A{}_C T_{(b)},{}^C{}_B - T_{(b)},{}^A{}_C T_{(a),}{}^C{}_B = f_a{}^c{}_b T_{(c)},{}^A{}_B. \tag{1.2}$$

These constants, $f_b{}^a{}_c$, have the following important properties:

$$
\begin{aligned}
f_a{}^m{}_b f_m{}^l{}_c + f_b{}^m{}_c f_m{}^l{}_a + f_c{}^m{}_a f_m{}^l{}_b = 0,\\
f_a{}^c{}_b = -f_b{}^c{}_a.
\end{aligned} \tag{1.3}
$$

The relations (1.3) can be easily obtained from Jacobi's identity and the definition (1.2).

Now from the invariant character of I under the transformation (1.1) and from the fact that this invariance is always preserved for an arbitrary domain Ω, we have the invariance of the Lagrangian density itself. Namely we have

$$
\delta L \equiv \frac{\partial L}{\partial Q^A}\delta Q^A + \frac{\partial L}{\partial Q^A{}_{,\mu}}\delta Q^A{}_{,\mu} \equiv 0. \tag{1.4}
$$

The symbol $\equiv$ means that δL must vanish at any world point and further that this relation does not depend on the behavior of Q^A and $Q^A{}_{,\mu}$. Substituting (1.1) into (1.4) we get

$$
\frac{\partial L}{\partial Q^A} T_{(a),}{}^A{}_B Q^B + \frac{\partial L}{\partial Q^A{}_{,\mu}} T_{(a),}{}^A{}_B Q^B{}_{,\mu} \equiv 0, \qquad (a = 1, 2, \cdots n) \tag{1.5}
$$

since the ϵ's are independent of each other. These n identities are the necessary and sufficient conditions for the invariance of I under G.

If we take into account the field equation for Q^A, we obtain from (1.5) the following n conservation laws:

$$
\partial J^\mu{}_a/\partial x^\mu = 0, \quad J^\mu{}_a = \frac{\partial L}{\partial Q^A{}_{,\mu}} T_{(a),}{}^A{}_B Q^B. \tag{1.6}
$$

This is so because (1.5) can be rewritten as follows:

$$
\left\{\frac{\partial L}{\partial Q^A} - \frac{\partial}{\partial x^\mu}\left(\frac{\partial L}{\partial Q^A{}_{,\mu}}\right)\right\}\delta Q^A + \frac{\partial}{\partial x^\mu}\left[\frac{\partial L}{\partial Q^A{}_{,\mu}}\delta Q^A\right] \equiv 0.
$$

The first term, $\{\quad\}\delta Q^A$, vanishes on account of the field equation.

Now let us consider the following transformation:

$$
\begin{aligned}
&\delta Q^A(x) = T_{(a),}{}^A{}_B \epsilon^a(x) Q^B,\\
&T_{(a),}{}^A{}_B = \text{constant},\\
&\quad \epsilon^a(x) = \text{infinitesimal arbitrary function},
\end{aligned} \tag{1.1$'$}
$$

instead of (1.1). In this case δL does not vanish but becomes

$$\delta L \equiv \{(1.5)\}_a \epsilon^a(x) + \frac{\partial L}{\partial Q^A{}_{,\mu}} T_{(a),}{}^A{}_B Q^B \frac{\partial \epsilon^a}{\partial x^\mu},$$

or

$$\delta L \equiv \frac{\partial L}{\partial Q^A{}_{,\mu}} T_{(a),}{}^A{}_B Q^B \frac{\partial \epsilon^a}{\partial x^\mu}, \tag{1.5$'$}$$

by virtue of the identity (1.5).

In order to preserve the invariance of the Lagrangian under (1.1)$'$, it is necessary to introduce a new field

$$A'^J(x), \quad J = 1, 2, \cdots M,$$

in such a way that the right-hand side of (1.5)$'$ can be cancelled with the contribution from this new field A'^J.

Now let us denote the new Lagrangian by

$$L'(Q^A, Q^A{}_{,\mu}, A'^J),$$

and consider the following transformation:

$$\begin{aligned} \delta Q^A &= T_{(a),}{}^A{}_B Q^B \epsilon^a(x), \\ \delta A'^J &= U_{(a)}{}^J{}_K A'^K \epsilon^a(x) + C^J{}_,{}^\mu{}_a \frac{\partial \epsilon^a}{\partial x^\mu}, \end{aligned} \tag{1.7}$$

where the coefficients U and C are unknown constants which will be determined later. In addition, let us propose that the new action integral I' is invariant under the transformation (1.7).

Our problem is to answer the five questions listed in the Introduction.

From the invariance postulate we get the following identity:

$$\delta L'(x) \equiv \frac{\partial L'}{\partial Q^A} \delta Q^A + \frac{\partial L'}{\partial Q^A{}_{,\mu}} \delta Q^A{}_{,\mu} + \frac{\partial L'}{\partial A'^J} \delta A'^J \equiv 0.$$

Inserting (1.7) into the above and taking account of the arbitrariness of choosing ϵ^a and $\partial \epsilon^a / \partial x^\mu$, we see that each coefficient of ϵ and $\partial \epsilon / \partial x$ must vanish independently. Namely, we have the identities

$$\frac{\partial L'}{\partial Q^A} T_{(a),}{}^A{}_B Q^B + \frac{\partial L'}{\partial Q^A{}_{,\mu}} T_{(a),}{}^A{}_B Q^B{}_{,\mu} + \frac{\partial L'}{\partial A'^J} U_{(a)}{}^J{}_K, A'^K \equiv 0, \tag{1.8}$$

$$\frac{\partial L'}{\partial Q^A{}_{,\mu}} T_{(a),}{}^A{}_B Q^B + \frac{\partial L'}{\partial A'^J} C^J{}_{,}{}^\mu{}_a \equiv 0. \tag{1.9}$$

Now, in order to be able to determine uniquely the A'^J-dependence of L', the number of the components of A'^J should be equal to the number of Eqs. (1.9), namely,

$$M = 4n$$

should hold. In addition, the matrix $C^J{}_{,}{}^\mu{}_a$ must be nonsingular. Thus there is the inverse of C defined by

$$C^J{}_{,}{}^\mu{}_a C^{-1a}{}_{\mu,K} = \delta^J{}_K, \quad C^{-1a}{}_{\mu,J} C^J{}_{,}{}^\nu{}_b = \delta^a{}_b \delta^\nu{}_\mu.$$

Then (1.9) can be rewritten as

$$\frac{\partial L'}{\partial A^a{}_\mu} + \frac{\partial L'}{\partial Q^A{}_{,\mu}} T_{(a),}{}^A{}_B Q^B \equiv 0,$$

where we have put

$$A^a{}_\mu = C^{-1a}{}_{\mu,j} A'^J.$$

Thus A'^J should be contained in L' only through the combination

$$\nabla_\mu Q^A \equiv \frac{\partial Q^A}{\partial x^\mu} - T_{(a),}{}^A{}_B Q^B C^{-1a}{}_{\mu,J} A'^J,$$

or

$$\nabla_\mu Q^A \equiv \frac{\partial Q^A}{\partial x^\mu} - T_{(a),}{}^A{}_B Q^B A^a{}_\mu. \tag{1.10}$$

By using $A^a{}_\mu$ in place of A'^J, the transformation character of A turns into

$$\delta A^a{}_\mu = S_{(c)}{}^{a,}{}_{\mu,}{}^\nu{}_b A^b{}_\nu \epsilon^c(x) + \partial \epsilon^a / \partial x^\mu, \tag{1.11}$$

where

$$S_{(c)}{}^{a,}{}_{\mu,}{}^\nu{}_b = C^{-1a}{}_{\mu,J} U_{(c)}{}^J{}_K C^K{}_{,}{}^\nu{}_b.$$

Now the new Lagrangian must have the form

$$L'(Q^A, Q^A{}_{,\mu}, A^a{}_\mu) = L''(Q^A, \nabla_\mu Q^A).$$

Therefore we have the relations

$$\frac{\partial L'}{\partial Q^A} = \left.\frac{\partial L''}{\partial Q^A}\right|_{\nabla Q\,\mathrm{const}} - \left.\frac{\partial L''}{\partial \nabla_\mu Q^B}\right|_{Q=\mathrm{const}} T_{(a),}{}^B{}_A A^a{}_\mu,$$

$$\frac{\partial L'}{\partial Q^A{}_{,\mu}} = \left.\frac{\partial L''}{\partial \nabla_\mu Q^A}\right|_{Q=\mathrm{const}},$$

$$\frac{\partial L'}{\partial A'^J} = -\left.\frac{\partial L''}{\partial \nabla_\nu Q^A}\right|_{Q=\mathrm{const}} T_{(b),}{}^A{}_B Q^B C^{-1b}{}_{\nu,J}.$$

By using these relations, (1.8) becomes

$$\begin{aligned}
&\left.\frac{\partial L''}{\partial Q^A}\right|_{\nabla Q=\mathrm{const}} T_{(a),}{}^A{}_B Q^B \\
&\qquad + \left.\frac{\partial L''}{\partial \nabla_\mu Q^A}\right|_{Q=\mathrm{const}} T_{(a),}{}^A{}_B \nabla_\mu Q^B \\
&\qquad + \left.\frac{\partial L''}{\partial \nabla_\mu Q^A}\right|_{Q=\mathrm{const}} Q^B A^b{}_\nu \{[T_{(a)}\, T_{(b)}]^A{}_B \delta^\nu{}_\mu \\
&\qquad - S_{(a)}{}^d{}_{\mu,}{}^\nu{}_b T_{(d),}{}^A{}_B\} \equiv 0.
\end{aligned} \tag{1.12}$$

If we put[2]

$$L''(Q^A, \nabla_\mu Q^A) = L(Q^A, \nabla_\mu, Q^A),$$

namely, put L'' to be what is obtained by replacing $\partial Q^A/\partial x^\mu$ in the original Lagrangian L with the "covariant derivative," $\nabla_\mu Q^A$, then on account of the identity (1.5), the first and the second terms (1.12) cancel each other. The remaining terms of (1.12) can be rewritten in the following way owing to the group character (1.2):

$$\left.\frac{\partial L}{\partial \nabla_\mu Q^A}\right|_{Q=\mathrm{const}} Q^B A^b{}_\nu \{f_a{}^d{}_b \delta^\nu{}_\mu - S_{(a)}{}^{d,}{}_\mu{}^\nu{}_b\} T_{(d),}{}^A{}_B \equiv 0.$$

Therefore we can determine the unknown coefficient S as follows:

$$S_{(a)}{}^{c,}{}_{\mu,}{}^\nu{}_b = \delta^\nu{}_\mu f_a{}^c{}_b. \tag{1.13}$$

Using this expression for S, we can easily show the covariant character of the derivative $\nabla_\mu Q^A$, i.e.,

$$\delta \nabla_\mu Q^A = T_{(a),}{}^A{}_B \epsilon^a(x) \nabla_\mu Q^B. \tag{1.14}$$

[2]This particular choice of L'' is due to the requirement that when the field A is assumed to vanish, we must have the original Lagrangian L.

Now let us investigate the possible type of the Lagrangian for the free A-field. Let it be denoted by

$$L_0(A^a{}_\mu, A^a{}_{\mu,\nu}), \quad A^a{}_{\mu,\nu} = \partial A^a{}_\mu / \partial x^\nu.$$

The invariance postulate for L_0 under the transformation (1.11) leads to

$$\frac{\partial L_0}{\partial A^a{}_\mu} f_b{}^a{}_c A^c{}_\mu + \frac{\partial L_0}{\partial A^a{}_{\mu,\nu}} \equiv 0, \tag{1.15}$$

$$\frac{\partial L_0}{\partial A^a{}_\mu} + \frac{\partial L_0}{\partial A^b{}_{\nu,\mu}} f_a{}^c{}_c A^c{}_\nu \equiv 0, \tag{1.16}$$

$$\frac{\partial L_0}{\partial A^a{}_{\mu,\nu}} + \frac{\partial L_0}{\partial A^a{}_{\nu,\mu}} \equiv 0. \tag{1.17}$$

From (1.17) we see that the derivative of A should be contained in L_0 through the combination

$$A^a{}_{[\mu,\nu]} \equiv \frac{\partial}{\partial x^\mu} A^a{}_\nu - \frac{\partial}{\partial x^\nu} A^a{}_\nu.$$

Thus (1.16) can be written as

$$\frac{\partial L_0}{\partial A^a{}_\mu} \equiv \frac{\partial L_0}{\partial A^c{}_{[\nu,\mu]}} f_a{}^c{}_b A^b{}_\nu. \tag{1.16$'$}$$

(1.16)′ means that the derivative of A appears in L_0, only through the particular combination

$$F^a{}_{\mu\nu} = \frac{\partial A^a{}_\nu}{\partial x^\mu} - \frac{\partial A^a{}_\mu}{\partial x^\nu} - \tfrac{1}{2} f_b{}^a{}_c (A^b{}_\mu A^c{}_\nu - A^b{}_\nu A^c{}_\mu). \tag{1.18}$$

Finally, substituting (1.16) into the first term of (1.15), we get

$$\frac{1}{2}\frac{\partial L_0}{\partial F^a{}_{\nu\mu}} \{ f_c{}^a{}_b A^b{}_{[\nu,\mu]} + \tfrac{1}{2}(f_d{}^a{}_b f_c{}^d{}_e - f_d{}^a{}_e f_c{}^d{}_b)(A^b{}_\nu A^e{}_\mu - A^b{}_\mu A^e{}_\nu)\} \equiv 0,$$

or by virtue of (1.3) we have

$$\frac{1}{2}\frac{\partial L_0}{\partial F^a{}_{\mu\nu}} f_c{}^a{}_b F^b{}_{\mu\nu} \equiv 0. \tag{1.19}$$

(See Appendix I.) Since L_0 must have the form

$$L_0(A, \partial A/\partial x) \equiv L_0'(A^a{}_\mu, F^a{}_{\mu\nu}),$$

we have the relations

$$\left.\frac{\partial L_0}{\partial A^a{}_{\nu\mu}}\right|_{A=\text{const}} = \left.\frac{\partial L_0{}'}{\partial F^a{}_{\mu\nu}}\right|_{A=\text{const},}$$

$$\left.\frac{\partial L_0}{\partial A^a{}_{\mu}}\right|_{\partial A/\partial x=\text{const}} = \left.\frac{\partial L_0{}'}{\partial A^a{}_{\mu}}\right|_{F=\text{const}} - \left.\frac{\partial L_0{}'}{\partial F^b{}_{\mu\nu}}\right|_{A=\text{const}} f_a{}^b{}_c A^c{}_\nu .$$

From these relations and (1.16), we have

$$\left.\frac{\partial L_0{}'}{\partial A^a{}_{\mu}}\right|_{F=\text{const}} \equiv 0.$$

Namely L_0 must be a function of F alone and must satisfy the identity (1.19).

As may easily be seen, the transformation character of F is given by

$$\delta F^a{}_{\mu\nu} = \epsilon^b(x) f_b{}^a{}_c F^c{}_{\mu\nu}. \tag{1.20}$$

Equation (1.20) can be verified by using the relation (1.3).

Now let us define a set of matrices, $M_{(1)}, M_{(2)}, \ldots M_{(n)}$, in the following way:

$$(a, b)\text{-element of } M_{(c)} \equiv M_{(c)}{}^a{}_b = f_c{}^a{}_b, \qquad (a, b, c = 1, 2, \cdots n).$$

Then these matrices are a representation of degree n for the generators of the Lie group G, since the relation (1.3) can be written as

$$[M_{(a)}, M_{(b)}]^l{}_c = f_a{}^m{}_b M_{(m),}{}^l{}_c .$$

Therefore (1.20) shows that n quantities, $F^1{}_{\mu\nu}, F^2{}_{\mu\nu}, \cdots F^n{}_{\mu\nu}$, are transformed cogradiently to the transformation of Q.

So far we have not used the field equations of A and Q.

The variation of the total Lagrangian density

$$L_T = L_0(F) + L(Q, \nabla Q)$$

can be rewritten as

$$\frac{\delta L_T}{\delta Q^A}\delta Q^A + \frac{\delta L_T}{\delta A^a{}_\mu} f_b{}^a{}_c \epsilon^b A^c{}_\mu - \frac{\partial}{\partial x^\mu}\left(\frac{\delta L_T}{\delta A^a{}_\mu}\right)\epsilon^a + \frac{\partial}{\partial x^\mu}\left\{\frac{\partial L}{\partial \nabla_\mu Q^A}\delta Q^A + \frac{\partial L_0}{\partial F^a{}_{\mu\nu}}\delta A^a{}_\nu + \frac{\delta L_T}{\delta A^a{}_\mu}\epsilon^a\right\} \equiv 0, \tag{1.21}$$

where the following abbreviations have been used:

$$\frac{\delta L_T}{\delta Q^A} = \frac{\partial L_T}{\partial Q^A} - \frac{\partial}{\partial x^\mu}\left(\frac{\partial L_T}{\partial Q^A{}_{,\mu}}\right), \qquad \frac{\delta L_T}{\delta A^a{}_\mu} = \frac{\partial L_T}{\partial A^a{}_\mu} - \frac{\partial}{\partial x^\nu}\left(\frac{\partial L_T}{\partial A^a{}_{\mu,\nu}}\right).$$

Now let us choose the arbitrary function $\epsilon^a(x)$ in such a way that the values of all the ϵ's and $\partial\epsilon/\partial x$'s vanish on the boundary surface of the integration domain Ω. Then the integration of (1.21) over the domain Ω becomes

$$\int_\Omega K d^4x \equiv 0, \tag{1.22}$$

with the abbreviation

$$K = \frac{\delta L_T}{\delta Q^A}\delta Q^A + \frac{\delta L_T}{\delta A^a{}_\mu} f_b{}^a{}_c \epsilon^b A^c{}_\mu - \frac{\partial}{\partial x^\mu}\left(\frac{\delta L_T}{\delta A^a{}_\mu}\right)\epsilon^a,$$

because the integration of the divergence term in (1.21) vanishes on account of our special choice of the ϵ's. Since the ϵ's can be chosen arbitrarily within Ω, K must vanish at every point in Ω, as is easily seen from (1.22).

Consequently the identity (1.21) are separated into the following two relations:

$$K \equiv 0, \tag{1.23}$$

and

$$\frac{\partial}{\partial x^\mu}\left\{\frac{\partial L}{\partial \nabla_\mu Q^A}\delta Q^A + \frac{\partial L_0}{\partial F^a{}_{\mu\nu}}\delta A^a{}_\nu + \frac{\delta L_T}{\delta A^a{}_\mu}\epsilon^a\right\} \equiv 0. \tag{1.24}$$

From (1.24), we have

$$\frac{\partial}{\partial x^\mu}\left\{\frac{\partial L}{\partial \nabla_\mu Q^A} T_{(a),}{}^A{}_B Q^B + \frac{\partial L_0}{\partial F^b{}_{\mu\nu}} f_a{}^b{}_c A^c{}_\nu + \frac{\delta L_T}{\delta A^a{}_\mu}\right\} \equiv 0, \tag{1.25}$$

$$\frac{\partial L}{\partial \nabla_\mu Q^A} T_{(a),}{}^A{}_B Q^B + \frac{\partial L_0}{\partial F^b{}_{\mu\nu}} f_a{}^b{}_c A^c{}_\nu + \frac{\partial L_T}{\partial A^a{}_\mu} \equiv 0, \tag{1.26}$$

and

$$\frac{\partial L_0}{\partial F^a{}_{\mu\nu}} + \frac{\partial L_0}{\partial F^a{}_{\nu\mu}} \equiv 0.$$

Put

$$J^\mu{}_a = \partial L_T / \partial A^a{}_\mu. \tag{1.27}$$

Then (1.26) leads to

$$J^{\mu}{}_{a} \equiv -\left(\frac{\partial L}{\partial \nabla_{\mu} Q^{A}} T_{(a),}{}^{A}{}_{B} Q^{B} + \frac{\partial L_{0}}{\partial F^{b}{}_{\mu\nu}} f_{a}{}^{b}{}_{c} A^{c}{}_{\nu}\right), \tag{1.28}$$

and (1.25) becomes

$$\frac{\partial J^{\mu}{}_{a}}{\partial x^{\mu}} \equiv \frac{\partial}{\partial x^{\mu}} \left\{\frac{\delta L_{T}}{\delta A^{a}{}_{\mu}}\right\}. \tag{1.29}$$

If we use the field equation

$$\delta L_{T}/\delta A^{a}{}_{\mu} = 0,$$

then we have the conservation of the "current," i.e.,

$$\partial J^{\mu}{}_{a}/\partial x^{\mu} = 0 \quad (a = 1, 2, \cdots n). \tag{1.30}$$

Thus we have obtained a general rule for introducing a new field A in a definite way when there exists some conservation law such as (1.6) or there is a Lie group depending upon some parameters under which the system is invariant.

In the following sections we shall consider the following groups as examples of the original Lie group: (1) the phase transformation of a charged field, (2) the rotation group in the isotopic spin space, and (3) the Lorentz group.

2. Phase Transformation Group and the Electromagnetic Field

Let us consider a charged field Q and Q^{*}. The Lagrangian of this system is assumed to be invariant under the phase transformation

$$\delta Q^{A} = i\alpha Q^{A}, \quad \delta Q^{A*} = -i\alpha Q^{A*}, \quad \alpha = \text{a real constant}.$$

Since this one-parameter group is commutative, the structure constant, of course, vanishes. By replacing the constant α with a scalar function $\lambda(x)$, a vector field $A_{\mu}(x)$ is introduced. The transformation character of $A_{\mu}(x)$ is given by

$$\delta A_{\mu} = \partial\lambda/\partial x^{\mu},$$

following the general formula (1.11). The new Lagrangian L' has the form

$$L' = L(Q, Q^{*}, \nabla_{\mu} Q, \nabla_{\mu} Q^{*})$$

where $\nabla_\mu Q$ and $\nabla_\mu Q^*$ are given by

$$\nabla_\mu Q^A = \frac{\partial Q^A}{\partial X^\mu} - iA_\mu Q^A, \qquad \nabla_\mu Q^{A*} = \frac{\partial Q^{A*}}{\partial x^\mu} + iA_\mu Q^{A*},$$

because in the present case

$$T^A{}_B = i\delta^A{}_B \quad \text{for} \quad Q^A, \quad T^A{}_B - = i\delta^A{}_B \quad \text{for} \quad Q^{A*}.$$

The Lagrangian L_0 for the free A_μ-field is

$$L_0 = L_0(F_{\mu\nu}),$$

where

$$F_{\mu\nu} = \frac{\partial A_\nu}{\partial x^\mu} - \frac{\partial A_\mu}{\partial x^\nu}.$$

The current J^μ can be obtained from the two different expressions

$$J^\mu = \frac{\partial L_T}{\partial A_\mu} = -i\left(\frac{\partial L}{\partial \nabla_\mu Q^A} Q^A - \frac{\partial L}{\partial \nabla_\mu Q^{A*}} Q^{A*}\right).$$

3. Rotation Group in Isotopic Spin Space and the Yang-Mills Field

As an example let us consider a system of proton and neutron fields;

$$\psi^\alpha = \begin{pmatrix} \psi_1 \\ \psi^2 \end{pmatrix} = \begin{pmatrix} \text{proton} \\ \text{neutron} \end{pmatrix}.$$

The Lagrangian in the charge-independent theory is invariant under the rotation in the three-dimensional isotopic spin space:

$$\delta\psi^\alpha = i\sum_{c=1}^{3} \epsilon^c \tau_{(c)}{}^\alpha{}_\beta \psi^\beta, \quad \delta\bar{\psi}_\alpha = -i\sum_{c=1}^{3} \epsilon^c \bar{\psi}_\beta \tau_{(c)}{}^\beta{}_\alpha, \tag{3.1}$$

where $\tau_{(1)}$, $\tau_{(2)}$, and $\tau_{(3)}$ are the usual isotopic spin matrices.

In this case the general notation T in Sec. 2 corresponds to τ as follows

$$T_{(c)},{}^{A}{}_{B} \to i\tau_{(c)},{}^{\alpha}{}_{\beta}, \quad \begin{pmatrix} \alpha, \beta = 1, 2 \\ c = 1, 2, 3 \end{pmatrix}.$$

By replacing the parameters, ϵ^a, with a set of functions, $\epsilon^a(x)$, the Yang-Mills field

$$B^c{}_\mu(x) \quad (c = 1, 2, 3)$$

is introduced, and this appears in the Lagrangian through the combination [see (1.10)]

$$\nabla_\mu \psi^\alpha = \partial\psi^\alpha/\partial x^\mu - i\tau_{(c)},{}^\alpha{}_\beta \psi^\beta B^c{}_\mu. \tag{3.2}$$

The variation of $B^c{}_\mu$ is given by [see (1.10) and (1.13)]

$$\delta B^c{}_\mu = f_a{}^c{}_b \epsilon^a(x) B^b{}_\mu + \frac{\partial \epsilon^c}{\partial x^\mu}, \tag{3.3}$$

where $f_a{}^c{}_b$ is defined by

$$[i\tau_{(a)}, i\tau_{(b)}] = f_a{}^c{}_b \cdot i\tau_{(c)}. \tag{3.4}$$

The derivative of $B^a{}_\mu$ can appear only through the combination [see (1.18)]

$$F^a{}_{\mu\nu} = \frac{\partial B^a{}_\nu}{\partial x^\mu} - \frac{\partial B^a{}_\mu}{\partial x^\nu} - \frac{1}{2} f_b{}^a{}_c (B^b{}_\mu B^c{}_\nu - B^b{}_\nu B^c{}_\mu). \tag{3.5}$$

The variations of $\nabla_\mu \psi$ and $F^a{}_{\mu\nu}$ are as follows:

$$\delta \nabla_\mu \psi^\alpha = i\epsilon^c \tau_{(c)},{}^\alpha{}_\beta \nabla_\mu \psi^\beta$$

and

$$\delta F^a{}_{\mu\nu} = \epsilon^c f_c{}^a{}_b F^b{}_{\mu\nu}. \tag{3.6}$$

As was stated in Sec. 1, $F^a{}_{\mu\nu}$ is transformed under the rotation group as a vector, namely, the isotopic spin of this B-field is unity. The expression for the "current" has the form [see (1.25) and (1.24)]:

$$J^\mu_c = \frac{\partial L_T}{\partial B^c{}_\mu} = -i \frac{\partial L}{\partial \nabla_\mu \psi^\alpha} \tau_{(c)},{}^\alpha{}_\beta \psi^\beta - \frac{\partial L_0}{F^a{}_{\mu\nu}} f_c{}^a{}_b B^b{}_\nu.$$

4. Lorentz Group and the Gravitational Field

Let us consider a system of fields $Q^A(x)$ being defined with respect to some Lorentz frame. In addition, let us assume that the action integral

$$I = \int L(Q^A, Q^A{}_{,k})d^4x$$

is invariant under any Lorentz transformation.

Now besides the x-system, let us introduce an arbitrary system of the curvilinear coordinates u^μ ($\mu = 1, 2, 3, 4$). In what follows, the Latin and Greek indices represent quantities defined with respect to the x-system (or the local Lorentz frame) and to the u-system respectively.

The square of the invariant length of the infinitesimal line element is given by

$$ds^2 = g^*{}_{ik}dx^i dx^k = g_{\mu\nu}du^\mu du^\nu,$$

where

$$g^*{}_{11} = g^*{}_{22} = g^*{}_{33} = -g^*{}_{44} = 1, \quad g^*{}_{ik} = 0 \quad \text{for} \quad i \neq k,$$

and

$$g_{\mu\nu}(u) = \frac{\partial x^i}{\partial u^\mu}\frac{\partial x^k}{\partial u^\nu} g^*{}_{ik}.$$

Let us introduce two sets of functions defined by

$$h^k{}_\mu(u) = \partial x^k / \partial u^\mu,$$

and

$$h_k{}^\mu(u) = \partial u^\mu / \partial x^k. \tag{4.1}$$

Then we have the following relations:

$$g^*{}_{kl}h^k{}_\mu h^l{}_\nu = g_{\mu\nu}(u), \quad G_{\mu\nu}h_k{}^\mu h_l{}^\nu = g^*{}_{kl}, \quad h_k{}^\mu h^l{}_\mu = \delta^1{}_k,$$
$$g^{kl*}h_k{}^\mu h_l{}^\nu = g^{\mu\nu}(u), \quad g^{\mu\nu}h^k{}_\mu h^l{}_\nu = g^{kl*}, \quad h_k{}^\mu h^k{}_\nu = \delta^\mu{}_\nu,$$
$$\det(g_{\mu\nu}) = g = -h^2 \equiv -[\det(h^k{}_\mu)]^2.$$

Raising or lowering of both kinds of suffices can be done by means of $g_{\mu\nu}$, $g^{\mu\nu}$ or $g^*{}_{kl}$. The geometrical meaning of the sets of $h^k{}_\mu$ and $h_k{}^\mu$, is obvious. The introduction of the four-world vector[3] $h_1{}^\mu$, $h_2{}^\mu$, $h_3{}^\mu$, and $h_4{}^\mu$ assigns

[3]The world vector means a vector which is defined with respect to the u-system.

respectively a local Lorentz frame to every world point. Of course, the local frames at every world point are transformed in the same way under any Lorentz transformation, i.e.,

$$x^k \to x^k + \epsilon^k{}_l x^l,$$
$$h_k{}^\mu \to h_k{}^\mu + \delta h_k{}^\mu, \quad \epsilon^{kl} = -\epsilon^{lk}$$
$$\delta h_k{}^\mu = -\epsilon^l{}_k h_l{}^\mu.$$

On account of this geometrical meaning of the h's, we can transform the world tensor into the corresponding local tensor defined with respect to the local frame, or vice versa, using $h_k{}^\mu$ or $h^k{}_\mu$. For example,

$$Q^k(u) = h^k{}_\mu(u) Q^\mu(u), \quad Q^\mu(u) = h_k{}^\mu(u) Q^k(u),$$

where the abbreviation

$$Q^k(u) = Q^k\{x(u)\}$$

has been used.

In this way we can rewrite the action integral as follows:

$$I = \int \mathfrak{L}(Q^A(u), Q^A{}_{,\mu}(u), h^k{}_\mu(u)) d^4u,$$

where $\mathfrak{L}$ is defined by

$$\mathfrak{L} = L(Q^A(u), h_k{}^\mu(u) Q^A{}_{,\mu}(u))h, \tag{4.2}$$

and $Q^A{}_{,\mu}$ stands for

$$\partial Q^A(u)/\partial u^\mu.$$

The reason for the fact that the Q^A in (4.2) is not transformed into the corresponding world quantity is that if Q^A is a spinor this rewriting is not possible, because the spinor can be well defined only with respect to a Lorentz frame.

Now I is invariant under the following two kinds of transformations[4,5]

(1) The Lorentz transformation

$$\delta h^k{}_\mu = \epsilon^k{}_l h^l{}_\mu,$$
$$\delta Q^A = \tfrac{1}{2} T_{(kl)},{}^A{}_B Q^B \epsilon^{kl},$$
$$u^\mu = \text{unchanged}, \tag{4.3}$$

[4]R. Utiyama, Progr. Theoret. Phys. (Japan) **2**, 38 (1947).

[5]L. Rosenfeld, Ann. Physik **5**, 113 (1930).

where $T_{(kl),}{}^{A}{}_{B}$ is the (A, B)-element of the $N \times N$ matrix $T_{(kl)}$ which is the representation of the generator of the Lorentz group. The matrix $T_{(kl)}$ satisfies the relation

$$[T_{(kl)}, T_{(m,n)}] = \tfrac{1}{2} f_{kl,}{}^{ab}{}_{mn} T_{(ab)}, \qquad T_{(kl)} = -T_{(lk)}.$$

(2) The general point transformation

$$\begin{aligned}
u^{\mu} &\rightarrow u^{\mu} + \lambda^{\mu}(u) = u'^{\mu}, \\
\lambda^{\mu}(u) &= \text{an arbitrary function of } u, \\
\delta h^{k}{}_{\mu} &= -\frac{\partial \lambda^{\nu}}{\partial u^{\nu}} h^{k}{}_{\nu}, \\
\delta Q^{A}(u) &\equiv Q^{A'}(u') - Q^{A}(u) = 0, \\
\delta Q^{A}{}_{,\mu} &= -\frac{\partial \lambda^{\nu}}{\partial u^{\mu}} Q^{A}{}_{,\nu}.
\end{aligned} \tag{4.4}$$

Now our Lagrangian (4.2) has the suitable form for the application of the general method stated in I, if the given functions, $h^{k}{}_{\mu}$, are regarded as a set of field quantities satisfying the condition:

$$\partial h^{k}{}_{\mu}/\partial u^{\nu} = \partial h^{k}{}_{\nu}/\partial u^{\mu}, \tag{4.5}$$

and having the transformation character (4.3) under the Lorentz group. Though we will omit the condition (4.5), the invariance of I under the transformations (4.3) and (4.4) still holds. The only role of (4.5) is to guarantee the possibility of finding the simplest and most convenient system of coordinates $(x^{1}, \cdots, x^{4})$. In fact if we replace the parameters, ϵ^{ik}, with a set of arbitrary functions, $\epsilon^{ik}(u)$, after the Lorentz transformation depending on such $\epsilon(u)$'s, the relation (4.5) is destroyed.

The condition (4.5) is inconsistent with the application of the general procedure of Sec. 1 to the present problem. Accordingly we shall consider the h's as a set of *16 independent given functions.*

Now following the prescription of Sec. 1, let us consider the "generalized Lorentz transformation" depending upon a set of arbitrary functions $\epsilon^{ik}(u)$ instead of the parameters ϵ^{ik}. Under this transformation, Q^{A} and $h^{k}{}_{\mu}$ are assumed to be transformed as

$$\begin{aligned}
\delta Q^{A} &= \tfrac{1}{2}\epsilon^{kl}(u) T_{(kl),}{}^{A}{}_{B} Q^{B}, \\
\delta h^{k}{}_{\mu} &= \epsilon^{k}{}_{l}(u) h^{l}{}_{\mu}.
\end{aligned} \tag{4.6}$$

Then, in order to retain the invariance of I under the transformation (4.6), it is necessary to introduce a new field

$$A^{kl}{}_{\mu}(u) = -A^{lk}{}_{\mu}(u),$$

which has the following transformation character according to (1.11):

$$\begin{aligned}\delta A^{kl}{}_{\mu} &= \tfrac{1}{4} f_{ab},{}^{kl}{}_{hg}\epsilon^{ab}(u) A^{hg}{}_{\mu} + \frac{\partial \epsilon^{kl}}{\partial u^{\mu}} \\ &= \epsilon^{k}{}_{m} A^{ml}{}_{\mu} + \epsilon^{l}{}_{m} A^{km}{}_{\mu} + \frac{\partial \epsilon^{kl}}{\partial u^{\mu}}. \end{aligned} \tag{4.7}$$

Furthermore the new Lagrangian is given by

$$\mathfrak{L}(Q^A, \nabla_{\mu} Q^A, h^k{}_{\mu}) = hL\{Q^A, (h_k{}^{\mu} \nabla_{\mu} Q^A)\} \tag{4.8}$$

where[6]

$$\nabla_{\mu} Q^A = \frac{\partial Q^A}{\partial u^{\mu}} - \frac{1}{2} A^{kl}{}_{\mu} T_{(kl),}{}^{A}{}_{B} Q^B. \tag{4.9}$$

[see (1.10)].

The factors $\frac{1}{2}$ in (4.9) and $\frac{1}{4}$ in (4.7) are necessary because in summing up the terms in these expressions with respect to the dummy suffices the same contributions are counted twice or four times.

Because of the "general Lorentz transformation," under which each local frame at each world point is transformed differently, the relation (4.5) was abandoned. Since this relation is satisfied only when the basic world is flat, we are forced to take as our basic space-time some Riemannian space with the metric

$$g_{\mu\nu}(u) = h^k{}_{\mu} h_{k\nu},$$

and the affine connection

$$\Gamma_{\mu\nu}{}^{\lambda} = \tfrac{1}{2} g^{\lambda\sigma} \left(\frac{\partial g_{\sigma\mu}}{\partial u^{\nu}} + \frac{\partial g_{\nu\sigma}}{\partial u^{\mu}} - \frac{\partial g_{\mu\nu}}{\partial u^{\sigma}} \right).$$

Accordingly we would expect that there exists some relationship between $A^{kl}{}_{\mu}$ and $h^k{}_{\mu}$.

In order to obtain this relationship let us consider, as an example, the local tensor

$$Q^{kl}(u)(= Q^A).$$

[6]F. J. Belinfante, Physica **7**, 305 (1940); K. Husimi, Proc. National Research Council of Japan **4**, 81 (1943).

Then from (4.9) we have

$$\nabla_\mu Q^{kl} = \frac{\partial Q^{kl}}{\partial u^\mu} - A^{km}{}_\mu Q_m{}^l - A^{lm}{}_\mu Q^k_m.$$

By using h, this can be rewritten as

$$\nabla_\mu Q^{k\nu} = \frac{\partial Q^{k\nu}}{\partial u^\mu} - A^{km}{}_\mu Q_m{}^\nu + \Gamma'{}_{\rho\mu}{}^\nu Q^{k\rho}, \tag{4.10}$$

where $Q^{m\nu}$ and Γ' are defined by

$$Q^{k\nu} = h_m{}^\nu Q^{km},$$

and

$$\Gamma'{}_{\nu\mu}{}^\rho = h_l{}^\rho \frac{\partial h^l_\nu}{\partial u^\mu} - h_k{}^\rho h_{l\nu} A^{kl}{}_\mu. \tag{4.11}$$

In general, the following relation is easily derived from (4.9):

$$\begin{aligned}\nabla_\mu Q^{kl\cdots,\rho\sigma\cdots}_{ab\cdots,\alpha\beta\cdots} = {} & \frac{\partial}{\partial u^\mu} Q^{kl\cdots,\rho\sigma\cdots}_{ab\cdots,\alpha\beta\cdots} - A^k{}_{i\mu} Q^{il\cdots,\rho\sigma\cdots}_{ab\cdots,\alpha\beta\cdots} - A^l{}_{i\mu} Q^{ki\cdots,\rho\sigma\cdots}_{ab\cdots,\alpha\beta\cdots} - \cdots \\ & + A^i{}_{a\mu} Q^{kl\cdots,\rho\sigma\cdots}_{ib\cdots,\alpha\beta\cdots} + A^i{}_{b\mu} Q^{kl\cdots,\rho\sigma\cdots}_{ai\cdots,\alpha\beta\cdots} + \cdots \\ & + \Gamma'{}_{\lambda\mu}{}^\rho Q^{kl\cdots,\lambda\sigma\cdots}_{ab\cdots,\alpha\beta\cdots} + \Gamma'{}_{\lambda\mu}{}^\sigma Q^{kl\cdots,\rho\lambda\cdots}_{ab\cdots,\alpha\beta\cdots} \\ & + \cdots - \Gamma'{}_{\alpha\mu}{}^\lambda Q^{kl\cdots,\rho\sigma\cdots}_{ab\cdots,\lambda\beta\cdots} - \Gamma'{}_{\beta\mu}{}^\lambda Q^{kl\cdots,\rho\sigma\cdots}_{ab\cdots,\alpha\lambda\cdots} - \cdots.\end{aligned} \tag{4.12}$$

This relation is nothing but the usual covariant derivative with the exception that for the Latin indices the $A^{kl}{}_{,\mu}$ must be inserted in place of the usual affinity Γ, and for the Greek indices our Γ' must be used instead of the Γ. Therefore for the world tensor $Q^{\mu\nu\cdots}$ our "covariant derivative" agrees with the usual covariant derivative with the affinity Γ'. Namely, if we use the symbol δ_μ for the usual covariant derivative, we get

$$\nabla_\mu Q^{\rho\nu} = \frac{\partial Q^{\rho\nu}}{\partial u^\mu} + \Gamma'{}_{\sigma\mu}{}^\rho Q^{\sigma\nu} + \Gamma'{}_{\sigma\mu}{}^\nu Q^{\rho\sigma} \equiv \delta_\mu Q^{\rho\nu}.$$

The relationship between A and h can be derived from the following consideration:

$$\begin{aligned}\nabla_\mu g^{kl*} &= -A^{kl}{}_\mu - A^{lk}{}_\mu = 0 \\ &= h^k{}_\rho h^l{}_\nu \nabla_\mu g^{\rho\nu} = h^k{}_\rho h^l{}_\nu \delta_\mu g^{\rho\nu}.\end{aligned}$$

From this expression we have

$$\delta_\mu g^{\rho\nu} = \frac{\partial g^{\rho\nu}}{\partial u^\mu} + \Gamma'_{\sigma\mu}{}^{\rho} g^{\sigma\nu} + \Gamma'_{\sigma\mu}{}^{\nu} g^{\rho\sigma} = 0. \tag{4.13}$$

If we assume that[7]

$$\Gamma'_{\mu\nu}{}^{\rho} = \Gamma'_{\nu\mu}{}^{\rho},$$

then we can solve (4.13) for Γ'. The solution is

$$\Gamma'_{\mu\nu}{}^{\rho} = \tfrac{1}{2} g^{\rho\sigma} \left(\frac{\partial g_{\sigma\mu}}{\partial u^\nu} + \frac{\partial g_{\nu\sigma}}{\partial u^\mu} - \frac{\partial g_{\mu\nu}}{\partial u^\sigma} \right) \equiv \Gamma_{\mu\nu}{}^{\rho},$$

or

$$h_l{}^{\rho} \frac{\partial h^l_\nu}{\partial u^\mu} - A^{\rho}{}_{\nu,\mu} = \Gamma_{\nu\mu}{}^{\rho}, \tag{4.14}$$

where

$$A^{\rho}{}_{\nu,\mu} = h_k{}^{\rho} h_{l\nu} a^{kl}{}_{,\mu}.$$

(4.14) is just the relation desired.

Now from (4.14) we see that $A^{\rho}{}_{\nu,\mu}$ is a world tensor of the third rank under the general point transformation (4.4) because the inhomogeneous term

$$-\frac{\partial^2 \lambda^\rho}{\partial u^\nu \partial u^\mu}$$

arising from $\delta\Gamma$ is cancelled out in virtue of the term $\delta(h\partial h/\partial u)$. Consequently the $A^{kl}{}_{\mu}$ is a covariant world vector under the u-transformation (4.4). Then it

[7] In general (4.13) gives the following expression for Γ', if the symmetry $\Gamma'_{\mu\nu}{}^{\lambda} = \Gamma'_{\nu\mu}{}^{\lambda}$ is not assumed:

$$\Gamma'_{\mu\nu}{}^{\rho} = \Gamma_{\mu,\nu}{}^{\rho} - (B^{\rho}{}_{\mu\nu} + B^{\rho}{}_{\nu,\mu}) + B_{\mu\nu,}{}^{\rho},$$

where Γ is the Christoffel affinity, while B is an arbitrary tensor with the symmetry character $B_{\mu\nu,\rho} = -B_{\nu\mu,\rho}$. [See E. Schrödinger, *Space-time Structure* (Cambridge University Press, Cambridge, 1950), p. 66.] Therefore our new field $A^{kl}{}_{\mu}$ (or $A_{\rho\nu,\mu} = h_{k\rho} h_{l\nu} A^{kl}{}_{\mu}$) can be represented in the following way in terms of h, $\partial h/\partial u$ and B:

$$A_{\rho\nu,\mu} = h_{l\rho} \frac{\partial h^l{}_\nu}{\partial u^\mu} - \Gamma_{\nu\mu,\rho} + B_{\rho\nu,\mu} + B_{\rho\mu,\nu} - B_{\nu\mu,\rho}.$$

The right-hand side of the above expression is antisymmetric in ρ and ν because of the fact that

$$\left(h_{l\rho} \frac{\partial h^l{}_\nu}{\partial u^\mu} - \Gamma_{\nu\mu,\rho} \right) + (\nu \text{ and } \rho \text{ interchanged}) = \frac{\partial g_{\rho\nu}}{\partial u^\mu} - \Gamma_{\nu\mu,\rho} - \Gamma_{\rho\mu,\nu} = 0.$$

Therefore the antisymmetry of $A_{\rho\nu,\mu}$ in ρ and ν does not give any restriction to the symmetry character of B. Now if B is assumed to vanish, we obtain the relation (4.14). On the other hand, if the basic space-time is flat, then A takes the form $A_{\rho\nu,\mu} = B_{\rho\nu,\mu} + B_{\rho\mu,\nu} - B_{\nu\mu,\rho}$, on account of the relation $\partial h^l{}_\nu/\partial u^\mu = \partial h^l{}_\mu/\partial u^\nu$, or, equivalently, $h_{l\rho}\partial h^l{}_\nu/\partial u^\mu = \Gamma_{\nu\mu,\rho}$.

is easily seen that our "covariant derivative" ∇_μ is in fact covariant under both kinds of transformations, namely, under any "general Lorentz transformation" and any u-transformation (4.4).

Thus we have obtained the general expression for the covariant derivative without using the concept of parallel displacement. For example, if the Q^A were the spinor field ψ, we have

$$\nabla_\mu \psi = \frac{\partial \psi}{\partial x^\mu} - \frac{i}{4} A^{kl}{}_\mu [\gamma_k, \gamma_l] \psi,$$

where the γ_k's are the usual Dirac γ matrices.

Now let us consider the Lagrangian $\mathfrak{L}_0$ of the free A-field:

$$\mathfrak{L}_0(h^k{}_\mu, A^{kl}{}_\mu, \partial A^{kl}{}_\mu / \partial u^\nu),$$

where the $h^k{}_\mu$ is necessary to raise or lower both kinds of tensor suffices.

From the invariance postulate for $\mathfrak{L}_0$ under the "general Lorentz transformation," we see that $\mathfrak{L}_0$ must have the form

$$\mathfrak{L}_0(h^k{}_\mu, F^{kl}{}_{\mu\nu}),$$

where F is defined by

$$\begin{aligned} F^{kl}{}_{\mu\nu} &= \frac{\partial A^{kl}{}_\nu}{\partial u^\mu} - \frac{\partial A^{kl}{}_\mu}{\partial u^\nu} - \frac{1}{4} f_{ab,}{}^{kl}{}_{mn} (A^{ab}{}_\mu A^{mn}{}_\nu - A^{ab}{}_\nu A^{mn}{}_\mu) \\ &= \frac{\partial A^{kl}{}_\nu}{\partial u^\mu} - \frac{\partial A^{kl}{}_\mu}{\partial u^\nu} + A^{kb}{}_\mu A^l{}_{b\nu} - A^{kb}{}_\nu A^l{}_{b\mu}. \end{aligned} \tag{4.15}$$

(4.15) can be rewritten formally as follows:

$$F^{kl}{}_{\mu\nu} = \nabla_\mu A^{kl}{}_\nu - \nabla_\nu A^{kl}{}_\mu - A^{kb}{}_\mu A^l{}_{b\nu} + A^{kb}{}_\nu A^l{}_{b\mu}, \tag{4.15$'$}$$

where $\nabla_\mu A^{kl}{}_\nu$ does not behave like a local tensor of the second rank, but is a covariant world tensor as the suffices μ and ν show. Using the expression (4.15)$'$ we can prove the following relation (see Appendix II)

$$F^{kl}{}_{\mu\nu} = h^{l\lambda} h^k{}_\alpha R^\alpha{}_{\lambda\mu\nu}, \tag{4.16}$$

where R is the Riemann-Christoffel curvature tensor:

$$R^\alpha{}_{\lambda\mu\nu} = \frac{\partial \Gamma_{\lambda\mu}{}^\alpha}{\partial u^\nu} - \frac{\partial \Gamma_{\lambda\nu}{}^\alpha}{\partial u^\mu} + \Gamma_{\lambda\mu}{}^\beta \Gamma_{\beta\nu}{}^\alpha - \Gamma_{\lambda\nu}{}^\beta \Gamma_{\beta\mu}{}^\alpha. \tag{4.17}$$

Though $h^k{}_\mu$ is contained in our Lagrangian as well as A and $\partial A/\partial u$, we can still prove that $A^{kl}{}_\mu$ appears in $\mathfrak{L}_0$ only through the combination F.

So far we have assumed that $h^k{}_\mu$ is a given function. The behavior of $h^k{}_\mu$ in general relativity is defined by the field equation derived by the variational principle.

The total Lagrangian density is now given by

$$\mathfrak{L}_T(Q^A, \nabla_\mu Q^A, h^k{}_\mu, \partial h^k{}_\mu/\partial u^\nu, \partial^2 h^k{}_\mu/\partial u^\nu \partial u^\lambda)$$
$$= \mathfrak{L}(Q^A, \nabla_\mu Q^A, h^k{}_\mu) + \mathfrak{L}_0(h^k{}_\mu, F^{kl}{}_{\mu\nu}).$$

The field equations for Q and h are[8]

$$\delta\mathfrak{L}/\delta Q^A = 0$$

and

$$\frac{\delta\mathfrak{L}_T}{\delta h^i{}_\mu} = \frac{\partial\mathfrak{L}_T}{\partial h^i{}_\mu} - \frac{\partial}{\partial u^\nu}\left(\frac{\partial\mathfrak{L}_T}{\partial h^i{}_{\mu,\nu}}\right) + \frac{\partial^2}{\partial u^\nu \partial u^\lambda}\left(\frac{\partial\mathfrak{L}_T}{\partial h^i{}_{\mu,\nu\lambda}}\right) - 0,$$

with the abbreviations

$$h^i{}_{\mu,\nu} = \partial h^i{}_\mu/\partial u^\nu, \quad h^i{}_{\mu,\nu\lambda} = \partial^2 h^i{}_\mu/\partial u^\nu \partial u^\lambda.$$

Now corresponding to (1.23) and (1.24) we have the following identities:

$$\frac{\delta\mathfrak{L}_T}{\delta Q^A}\delta Q^A + \frac{\delta\mathfrak{L}_T}{\delta h^i{}_\mu}\delta h^i{}_\mu \equiv 0, \tag{4.18}$$

and

$$(\partial/\partial u^\mu)\mathfrak{M}^\mu \equiv 0, \tag{4.19}$$

[8]If the $B_{\mu\nu,\rho}$ is taken into account, the Riemann-Christoffel tensor $R^\rho{}_{\lambda\mu\nu}$ is the same function as (4.17) but Γ' must be inserted in place of Γ in (4.17). Here Γ' is the affinity given in reference 7. In this case, in addition to the equations for the Q and h fields, we have the following equation:

$$\frac{\delta\mathfrak{L}_T}{\delta B_{\mu\nu,\lambda}} = \frac{1}{2}\frac{\partial\mathfrak{L}}{\partial A^{kl}{}_\rho}\frac{\partial A^{kl}{}_\rho}{\partial B_{\mu\nu,\lambda}} + \frac{1}{4}\frac{\partial\mathfrak{L}_0}{\partial F^{kl}{}_{\alpha\beta}}\frac{\partial F^{kl}{}_{\alpha\beta}}{\partial B_{\mu\nu,\lambda}} - \frac{1}{4}\frac{\partial}{\partial u^\rho}\left\{\frac{\partial\mathfrak{L}_0}{\partial F^{kl}{}_{\alpha\beta}}\frac{\partial F^{kl}{}_{\alpha\beta}}{\partial\left(\frac{\partial B_{\mu\nu,\lambda}}{\partial u^\rho}\right)}\right\} = 0.$$

Of course, in all these equations for the Q, h, and B fields, the affinity Γ' must be used instead of Γ. (4.18) and (4.19) also hold in this case, since B is invariant under the "general Lorentz transformations." Consequently this B field is of no use in avoiding the trivial result: $\mathfrak{M}^\mu{}_{ik} \equiv 0$.

where $\mathfrak{M}^\mu$ is

$$\mathfrak{M}^\mu = \frac{\partial \mathfrak{L}}{\partial \nabla_\mu Q^A} \delta Q^A + \frac{\partial \mathfrak{L}_T}{\partial h^i{}_{\rho,\mu}} \delta h^i{}_\rho + \frac{\partial \mathfrak{L}_T}{\partial h^i{}_{\rho,\mu\nu}} \delta h^i{}_{\rho,\nu} - \frac{\partial}{\partial u^\nu} \left(\frac{\partial \mathfrak{L}_T}{\partial h^i{}_{\rho,\mu\nu}} \right) \cdot \delta h^i{}_\rho. \tag{4.20}$$

Accordingly, the coefficient of $\partial^2 \epsilon / \partial u^2$ in (4.19) identically vanishes:

$$\frac{\partial \mathfrak{L}_T}{\partial h^i{}_{\rho,\mu\nu}} h_{k\rho} - \frac{\partial \mathfrak{L}_T}{\partial h^k{}_{\rho,\mu\nu}} h_{i\rho} \equiv 0.$$

Thus $\mathfrak{M}^\mu$ becomes

$$\mathfrak{M}^\mu = \tfrac{1}{2} \epsilon^{ik} \mathfrak{M}^\mu{}_{ik},$$

$$\mathfrak{M}^\mu{}_{ik} = \frac{\partial \mathfrak{L}}{\partial \nabla_\mu Q^A} T_{(ik),}{}^A{}_B Q^B + \left[\left\{ \frac{\partial \mathfrak{L}_T}{\partial h^i{}_{\rho,\mu}} h_{k\rho} + \frac{\partial \mathfrak{L}_T}{\partial h^i{}_{\rho,\mu\nu}} h_{k\rho,\nu} - \frac{\partial}{\partial u^\nu} \left(\frac{\partial \mathfrak{L}_T}{\partial h^i{}_{\rho,\mu\nu}} \right) h_{k\rho} \right\} - \{k \text{ and } i \text{ interchanged}\} \right].$$

Inserting this expression into (4.19), we have the trivial result

$$\partial \mathfrak{M}^\mu{}_{ik} / \partial u^\mu \equiv 0, \tag{4.21}$$

as

$$\mathfrak{M}^\mu{}_{ik} \equiv 0.$$

Since the Eulerian derivative $\delta L / \delta A$ appeared in (1.24), the nonvanishing "current" could be derived in the general theory in Sec. 1. In the present case, however, we have no Eulerian derivatives in $\mathfrak{M}$. Thus the field equations do not play any role in deriving the "current."

Now the usual equation for the gravitational field is derived by taking a particular Lagrangian

$$\mathfrak{L}_0 = hR,$$

where R is defined as follows:

$$R = g^{\mu\nu} R_{\mu\nu} = h_l{}^\mu h_k{}^\nu F^{kl}{}_{\mu\nu}, \quad R_{\mu\nu} = R^\lambda{}_{\mu\nu\lambda}.$$

Taking the variation with regard to h, we get

$$\frac{\delta \mathfrak{L}_0}{\delta h^i{}_\mu} + \frac{\delta \mathfrak{L}}{\delta h^i{}_\mu} = 0.$$

The following relation is now easily verified:

$$\frac{\delta \mathfrak{L}_0}{\delta h^i{}_\mu}\delta h^i{}_\mu = \frac{\delta \mathfrak{L}_0}{\delta g_{\rho\mu}} h_{i\rho}\delta h^i{}_\mu + \frac{\partial}{\partial u^\mu}\left\{\frac{\partial \mathfrak{L}_0}{\partial g_{\rho\sigma,\mu}} h_{i\rho}\delta h^i{}_\sigma\right\},$$

where[9]

$$\delta \mathfrak{L}_0/\delta g_{\rho\sigma} = -h(R^{\rho\sigma} - \tfrac{1}{2}g^{\rho\sigma}R) = -\mathfrak{G}^{\rho\sigma}.$$

Thus we have

$$h_{i\sigma}\mathfrak{G}^{\rho\sigma} = -\mathfrak{I}^\rho{}_i,$$

with

$$\mathfrak{I}^\rho{}_i = -\delta\mathfrak{L}/\delta h^i{}_\rho,$$

or we have

$$\mathfrak{G}^{\rho\sigma} = -\mathfrak{I}^{\rho\sigma}.$$

Here $\mathfrak{I}$ is the symmetric energy-momentum-tensor density of the original field Q. The symmetry character of $\mathfrak{I}$ can be proved in the following way.

Since the Lagrangian $\mathfrak{L}$ for the Q-field is also invariant under the "general Lorentz transformation," we have an identity similar to (4.18):

$$\frac{\delta \mathfrak{L}}{\delta Q^A}\delta Q^A + \frac{\delta \mathfrak{L}}{\delta h^i{}_\mu}\delta h^i{}_\mu \equiv 0.$$

Inserting the field equation for the Q-field into this identity, we get

$$\mathfrak{I}_{ik} = \frac{\delta \mathfrak{L}}{\delta h^i{}_\mu} h_{k\mu} = \frac{\delta \mathfrak{L}}{\delta h^k{}_\mu} h_{i\mu} = \mathfrak{I}_{ki}.$$

From this relation,

$$\mathfrak{I}^{\mu\rho} = h^{i\mu}h^{k\rho}\mathfrak{I}_{ik} = \mathfrak{I}^{\rho\mu}$$

can be easily derived.

[9] W. Pauli, Encyklopaedie der Mathematischen Wissenschaften (B. G. Teubner, Leipzig, 1904–1922), Vol. 5, Chap. 19, p. 621.

Acknowledgments

The author is most grateful to the Institute for Advanced Study for grant-in-aid and to Professor Robert Oppenheimer for the kind hospitality extended him there. He is also indebted to members of the Institute, especially to Dr. R. Arnowitt, for helpful conversations.

Appendix I. Condition (1.19)

Here we shall show how to construct an invariant in terms of $F^a{}_{\mu\nu}$.

Consider a quantity G_a, the transformation character of which is contragradient to that of $F^a{}_{\mu\nu}$ under the transformation (1.20).

Since

$$G_a F^a{}_{\mu\nu}$$

is invariant by definition, δG_a is given by

$$\delta G_A = -\epsilon^c f_c{}^b{}_a G_b.$$

For example,

$$K_{a,\mu\nu} = f_a{}^l{}_{m,} f_l{}^m{}_b F^b{}_{\mu\nu}$$

is transformed contragradiently to $F^a{}_{\mu\nu}$, for $\delta K_{a,\mu\nu}$ is

$$\begin{aligned}\delta K_{a,\mu\nu} &= f_a{}^l{}_m f_l{}^m{}_b \delta F^b{}_{\mu\nu}\\ &= f_a{}^l{}_m f_l{}^m{}_b f_j{}^b{}_k F^k{}_{\mu\nu} \epsilon^j.\end{aligned}$$

By taking into account the relation (1.3), the above expression becomes

$$\delta K_{a,\mu\nu} = f_a{}^l{}_m (f_k{}^b{}_l f_b{}^m{}_j + f_l{}^b{}_j f_b{}^m{}_k) F^k{}_{\mu\nu} \epsilon^j.$$

Using (1.3) again, we can rewrite the first term of this expression as follows:

$$\delta K_{a,\mu\nu} = f_k{}^b{}_l (f_j{}^m{}_a f_m{}^l{}_b + f_a{}^m{}_b f_m{}^l{}_j) F^k{}_{\mu\nu} \epsilon^j - f_k{}^m{}_b f_a{}^l{}_m f_l{}^b{}_j F^k{}_{\mu\nu} \epsilon^j.$$

Since the second term is cancelled with the last term, we have

$$\delta K_{a,\mu\nu} = -\epsilon^j f_j{}^m{}_a f_m{}^l{}_b f_l{}^b{}_k F^k{}_{\mu\nu} F^k_{\mu\nu} = -\epsilon^j f_j{}^m{}_a K_{m,\mu\nu}.$$

Now let us call $F^a{}_{\mu\nu}$ a contravariant vector, and $K_{a,\mu\nu}$ a covariant vector with regard to the transformation (1.20). In addition, let us propose that $f_b{}^a{}_c$ is contravariant with respect to the suffix a and covariant with respect to b and c. Then we see that $f_b{}^a{}_c$ is a constant and invariant tensor owing to the following fact that

$$\delta f_b{}^a{}_c = \epsilon^j (f_j{}^a{}_k f_b{}^k{}_c - f_j{}^k{}_b f_k{}^a{}_c - f_j{}^k{}_c f_b{}^a{}_k),$$

which vanishes in virtue of the relation (1.3). Hence this proposal concerning the transformation character of $f_b{}^a{}_c$ is compatible with the covariant character of $K_{a,\mu\nu}$.

Using the quantity

$$g_{ab} = f_a{}^l{}_m f_l{}^m{}_b = g_{ba}$$

and its inverse g^{ab}, we can easily construct a tensor algebra similar to that used in the theory of relativity. For example, we have the invariants

$$H_{\mu\nu,\rho\sigma} = g_{ab} F^a{}_{\mu\nu} F^b{}_{\rho\sigma} = H_{\rho\sigma,\mu\nu}.$$

In the case of the rotation group in three-dimensional isotopic spin space (see Sec. 3), $f_b{}^a{}_c$ has the following values:

$$\begin{cases} f_1{}^3{}_2 = f_3{}^2{}_1 = f_2{}^1{}_3 = -1, \\ f_i{}^l{}_k = -f_k{}^l{}_i, \\ \text{otherwise } f = 0. \end{cases}$$

Therefore, we have

$$g_{ab} = 2\delta_{ab},$$

and

$$H_{\mu\nu,\rho\sigma} = 2\delta_{ab} F^a{}_{\mu\nu} F^b{}_{\rho\sigma}.$$

Another familiar example is the case of the Lorentz group. Here we have

$$\tfrac{1}{4} f_{jk,}{}^{ab}{}_{cd} f_{ab,}{}^{cd}{}_{lm} = g^*{}_{jl} g^*{}_{km} - g^*{}_{jm} g^*{}_{kl},$$

and

$$\begin{aligned} H_{\mu\nu,\rho\sigma} &= \frac{1}{2^4} F^{jk}{}_{\mu\nu} f_{jk,}{}^{ab}{}_{cd} f_{ab,}{}^{cd}{}_{lm} F^{lm}{}_{\rho\sigma} \\ &= \tfrac{1}{2} F^{jk}{}_{\mu\nu} F^{lm}{}_{\rho\sigma} g^*{}_{jl} g^*{}_{km} = \tfrac{1}{2} F^{jk}{}_{\mu\nu} F_{jk,\rho\sigma}. \end{aligned}$$

If L_0 is a function of the invariant $H_{\mu\nu,\beta\sigma}$ alone, we can easily prove the identity

$$\frac{\partial L_0}{\partial F^a{}_{\mu\nu}} f_c{}^a{}_b F^b{}_{\mu\nu} \equiv 0. \tag{1.19}$$

Namely, the left-hand side can be written as

$$\frac{1}{4}\frac{\partial L_0}{\partial H_{\rho\sigma,\alpha\beta}}\frac{\partial H_{\rho\sigma,\alpha\beta}}{\partial F^a{}_{\mu\nu}} f_c{}^a{}_b F^b{}_{\mu\nu} = \frac{1}{4}\frac{\partial L_0}{\partial H_{\rho\sigma,\mu\nu}} F^d{}_{\rho\sigma} F^b{}_{\mu\nu}(g_{ad} f_c{}^a{}_b + g_{ab} f_c{}^a{}_d).$$

The factor in the bracket vanishes on account of the relation (1.3). Consequently there exists, in fact, a family of invariant Lagrangians, L_0, which are functions of $F^a{}_{\mu\nu}$ alone and satisfy the condition (1.19).

Appendix II. Proof of the Relation $F^{kl}{}_{\mu\nu} = h^{l\lambda} h^k{}_\alpha R^\alpha{}_{\lambda\mu\nu}$

$F^{kl}{}_{\mu\nu}$ is given by

$$F^{kl}{}_{\mu\nu} = \nabla_\mu A^{kl}{}_\nu - \nabla_\nu A^{kl}{}_\mu - A^{kb}{}_\mu A^l{}_{b\nu} + A^{kb}{}_\nu A^l{}_{b\mu}. \qquad (A.1)$$

Now according to the general rule (4.12)

$$\nabla_\mu A^{kl}{}_\nu = h^k{}_\rho h^l{}_\sigma \delta_\mu A^{\rho\sigma}{}_\nu, \qquad (A.2)$$

where δ_μ is the usual covariant derivative with the Christoffel affinity.

$A^{\rho\sigma}{}_{,\nu}$ is by virtue of (4.14) rewritten as

$$\begin{aligned} A^{\rho\sigma}{}_{,\nu} &= g^{\sigma\lambda} A^\rho{}_{\lambda,\nu} \\ &= g^{\sigma\lambda} h_l{}^\rho \left(\frac{\partial h^l{}_\lambda}{\partial u^\nu} - \Gamma_{\nu\lambda}{}^\alpha h^l{}_\alpha\right). \end{aligned}$$

If we suppose $h^l{}_\lambda$ as a simple covariant world vector and ignore the local suffix l, then the factor in the parenthesis in the above expression is just the usual covariant derivative of $h^l{}_\lambda$. Therefore we get

$$A^{\rho\sigma}{}_{,\nu} = g^{\sigma\lambda} h_l{}^\rho \delta_\nu h^l{}_\lambda.$$

On the other hand from (4.14) we have the relation

$$\delta_\nu h^l{}_\lambda = A^l{}_{m,\nu} h^m{}_\lambda. \qquad (A.3)$$

Thus we have

$$\delta_\mu A^{\rho\sigma}{}_{,\nu} = g^{\sigma\lambda} \delta_\mu (h_l{}^\rho \delta_\nu h^l{}_\lambda).$$

By using (A.3) this becomes

$$\delta_\mu A^{\rho\sigma}{}_{,\nu} = g^{\sigma\lambda} h_l{}^\rho \delta_\mu \delta_\nu h^l{}_\lambda - A^m{}_{l,\mu} A^l{}_{k,\nu} h_m{}^\rho h^{k\sigma}.$$

Inserting this expression into (A.1) with (A.2), we have

$$F^{kl}{}_{\mu\nu} = h^{l\lambda}(\delta_\mu \delta_\nu - \delta_\nu \delta_\mu) h^k{}_\lambda.$$

As is well known, the Riemann-Christoffel tensor is defined by

$$(\delta_\mu \delta_\nu - \delta_\nu \delta_\mu) V^\lambda = R^\rho{}_{\lambda\mu\nu} V_\rho$$

for an arbitrary covariant vector V_ρ.

Thus we get

$$F^{kl}{}_{\mu\nu} = h^{l\lambda} h^k{}_\alpha R^\alpha{}_{\lambda\mu\nu}.$$

Conclusion

As is well known, the advent of non-abelian gauge theory did not herald a new dawn for the nuclear interactions. The fact that a self-consistent theory had been formulated did not change the fact that the nuclear interactions were short-range and therefore precluded massless particles. Furthermore, all the indications were that the strong and weak interactions were mediated by pseudo-scalar and four-fermi interactions, respectively. Thus in 1954 the Yang-Mills proposal seemed destined to become no more than a theoretical toy. Only when the vector $(V - A)$ character of the weak currents emerged in 1958 was it possible to think seriously of applying the non-abelian theory to the real world.

Even then, there were formidable hurdles to be overcome. On the theoretical side was the problem that the theory did not appear to be renormalizable and that the infra-red problem for non-abelian theories was much more serious than for theories of electromagnetism. The phenomenological situation was even worse. The gauge fields appeared to be massless, in disagreement with the extremely short range of the nuclear interactions. Furthermore, the strong interactions appeared to be mediated by the pseudo-scalar π-mesons rather than vector-particles. Even for the weak interactions, which had a vector-like structure, the theory required that the charged currents be accompanied by a neutral one, for which there was absolutely no experimental evidence.

It is not surprising that the resolution of these problems took some time. What is astonishing is that they could be resolved at all. The first major advance was in connection with the mass problem, which was solved a decade later by the invention (or, more precisely, the importation from solid-state physics) of the spontaneous symmetry-breaking mechanism [1]. It was followed by the resolution of the renormalization problem through the use of gauge-invariant renormalization techniques [2], notably, dimensional regularization.

These two theoretical successes spurred a serious experimental search for the neutral weak current, and its discovery led to an extensive investigation into the low-energy structure of the weak interactions. The results turned out to be in complete agreement with gauge theory, in the form of the standard gauge-theory model [3], which was based on the Lie algebra $SU(2) \times U(1)$. Finally, in 1981 the matter was put beyond doubt by the experimental production [4] of the gauge-field particles predicted by that model, the well-known Z^o and

$W^{\pm}$ mesons. Experiments since then have been in such good agreement with the standard model that at present the problem [5] is to find a discrepancy that would point toward new physics!

Meanwhile, the situation with regard to the strong interactions was changing with the gradual realization that the nucleons and mesons were not elementary but composite particles, built of quarks [6]. This implied that the interactions between them were non-local and therefore unlikely to be fundamental. Then, in 1974, came the dramatic discovery that non-abelian gauge theories are asymptotically free [7], i.e., the strength of their coupling decreases with energy. The fact that asymptotic freedom agreed with high-energy observations and that its complement, infra-red slavery, would account for quark-confinement, strongly suggested that the strong interactions were described by a non-abelian gauge theory at a more fundamental level. The problem of finding the corresponding gauge group was conveniently solved by the existence of an internal unbroken continuous symmetry group, namely, the $SU(3)$ color group of the quarks, which had been introduced earlier for quite different (spin-statistical) reasons. Thus was born the present picture of the strong interactions in which the quarks interact by coupling locally to $SU(3)$ gauge fields (colored gluons) and thereby induce the observed non-local interactions of their baryon and meson composites. The confinement of the quarks and gluons means, of course, that it is difficult to find direct evidence for this theory, just as it would be difficult to find direct evidence for atomic theory using only molecular interactions. But the indirect evidence for $SU(3)$ gauge theory is now so strong that it is almost universally accepted [8].

In principle, the only difference between the gluonic interactions of the quarks and the electromagnetic interactions of the electrons and protons in atomic physics is that the gluons interact with each other because of the non-abelian nature of the group. But in practice this difference is of the utmost importance. It is the self-interaction of the gluons that is responsible for asymptotic freedom (small coupling at short distances) and, together with the long-range of the gluonic interactions, is supposed to be responsible for quark confinement (strong coupling at large distances). If the supposition is correct, it means that confinement is due partly to the fact that the gluons are massless. Thus, ironically, the infra-red problem that so embarrassed the early inventors of the theory, and led to Pauli's clash with Yang in 1954 (see Chapter 8), may turn out to be one of the theory's greatest assets!

From the above brief summary it will be clear that the developments in gauge theory subsequent to 1954 were just as complicated and fascinating as those prior to that date. But their proper recounting is beyond the scope of this book.

In conclusion, it might be of interest to recall a remark of Einstein concerning his celebrated gravitational equations,

$$G_{\mu\nu} = \kappa T_{\mu\nu}, \tag{172}$$

namely, that the left-hand side was made of ivory and the right-hand side of wood, meaning that the left-hand side was geometrical but the right-hand side was empirical and so the whole equation would eventually have to be motivated in a more profound manner, with the right-hand side incorporated into the left-hand one. (This has actually happened in some supersymmetric models.) The field equations for the gauge fields are in a similar situation since they take the form

$$D^{\nu} F_{\mu\nu} = J_{\mu}, \tag{173}$$

where the left-hand side is geometrical but the matter currents on the right-hand side, made of fermionic and scalar fields, are empirical. In this case, however, the problem of finding a universal theory may be even more complicated on account of the lack of universality in the gauge-field couplings.

The elevation of the gauge fields to the level of the gravitational field is a substantial achievement, but is by no means the end of the story. Indeed, there are two major limitations on the power of gauge theory. First, in its present form at least, it does not unify the fundamental interactions in an intrinsic way, in the sense that the coupling constants for the different interactions remain theoretically undetermined. Even in the standard electroweak model, the electromagnetic and weak coupling constants remain undetermined. Furthermore, gravitation is not unified with the other interactions, so the radiation fields still stand on the right-hand side of (172). Second, gauge theory provides no answers for the questions that arise concerning the matter fields, such as the distinction between baryons and leptons, the existence of three generations [9] of quark-lepton pairs, the origin and structure of the symmetry-breaking scalar sector, and of the quark mixing matrix [3][10].

What gauge theory does is to determine the interaction of the radiation fields with themselves and with matter, and to provide a universal and reliable framework for further investigation and development. All the major advances of recent years, such as grand unification, supersymmetry and string theory, have been made within this framework.

References

INTRODUCTION

[1] H. Minkowski, Phys. Zeitschr. **10** (1909) 104.
[2] Quoted in C. Lanczos, *Einstein and the Cosmic World Order*, Wiley, New York 1965.
[3] For a parallel discussion of unified theories during this period, see: V. Vizgen, *Unified Field Theories in the First Third of the Twentieth Century*, BirkHauser-Verlag 1994.
[4] D. Bailin, *Weak Interactions* Hilger, Bristol 1982. C. Lai, *Gauge Theory of the Weak and Electromagnetic Interactions*, World Scientific, Singapore 1981.
[5] P. Becher, P. Bohm, and H. Joos, *Gauge Theories of the Strong and Electroweak Interactions*, Wiley, New York 1984.
[6] M. Peshkin and A. Tonomura, *The Aharonov-Bohm Effect*, Lecture Notes in Physics **340**, Springer-Verlag 1989.
[7] C. Ehresmann, *Colloque de Topologie* Brussels 1950 pp. 29–55. S. Kobayashi and K. Nomizu, *Foundations of Differential Geometry*, Interscience, 1969.
[8] L. Okun, Proc. JINR-CERN School of Physics, CERN, Geneva 1983.
[9] C. Jarlskog, in Physica Scripta **24** No. 5. (1981) 867.
[10] A. Koestler, *The Sleepwalkers*, Grosset and Dunlap, New York 1963.

CHAPTER 1

[1] T. T. Wu and C. N. Yang, Phys. Rev. **D12** (1975) 3845.
[2] See, for example, T.-P. Chen and L.-F. Li, *Gauge Theory of Elementary Particle Physics*, Clarendon Press, Oxford (1984).
[3] F. J. Dyson, Amer. J. Phys. **58**(3) (1990) 209.
[4] E. Noether, Nachr. d. Kgl. Gessch. d. Wiss. Gottingen, Kl. Math-Phys. (1918) 235.
[5] W. Pauli, *General Principles of Quantum Mechanics*, Springer-Verlag 1980 (translated from *Wellenmechanik* in Handbuch der Physik 1933).
[6] S. Pokorski, *Gauge-Field Theories*, Cambridge Univ. Press 1987.
[7] R. Jackiw et al., *Current Algebra and Anomalies*, World Scientific, Singapore 1985.
[8] W. Pauli, *Theory of Relativity*, Pergamon, New York 1958.
[9] N. Straumann, *Albert Einstein: auf dem Weg zur Gravitationstheorie*. Viert. jr. schr. der Naturf. Ges. Zürich **139/3** (1994) 103.

CHAPTER 2

[1] C. Levi-Civita, Rend. Circ. Mat. Palermo **42** (1917).
[2] E. Cartan, C. R. Acad. Sci. **174** (1922) 593. See also [1] of Chapter 3.
[3] H. Weyl, Math. Zeit. **2** (1918) 384.
[4] H. Weyl, *Space-Time-Matter* (Raum-Zeit-Materie), Methuen, London (1922).
[5] E. Schrodinger, *Space-Time Structure*, Cambridge Univ. Press 1953.

CHAPTER 3

[1] E. Cartan, *On Manifolds with an Affine Connection and the Theory of General Relativity*, trans. A. Magnon and A. Ashtekar, Bibliopolis, Napoli 1986.
[2] F. Hehl et al., Rev. Mod. Phys. **48** (3) (1976) 393 and Proc. 1995 Erice School on Cosmology and Gravitation, World Scientific (1996).
[3] E. Schrodinger, Ann. d. Physik **79, 80, 81** (1926).
[4] T. Appelquist et al., *Modern Kaluza-Klein Theories*, Addison-Wesley, New York 1987. Y. Kubyshin et al., *Dimensional Reduction of Gauge Theories*, Lecture Notes in Physics, **349**, Springer-Verlag 1989. M. Duff et al. Physics Rep. **130C** (1986) 2.
[5] M. Green, J. Schwarz, and E. Witten, *Superstring Theory*, Cambridge Univ. Press, 1987.

CHAPTER 4

[1] V. Raman and P. Forman, Hist. Studies Phys. Sci. **1** (1969) 291.
[2] C. N. Yang, in *Schrodinger*, ed. C. Kilmister, Cambridge Univ. Press 1987.
[3] W. Gordon, Zeit. f. Physik, **40** (1926) 117: O. Klein, Zeit. f. Physik, **41** (1927) 407.
[4] P. Dirac, Proc. Roy. Soc. **A117** (1928) 610, **A118** (1928) 351.
[5] C. Lanczos (private communication). For a slightly different formulation, see J. Levy-Leblond, Comm. Math. Phys. **6** (1967) 286.
[6] C. N. Yang, in "Hermann Weyl's Contribution to Physics," in *Hermann Weyl*, ed. K. Chandrasekharan, Springer-Verlag 1980.
[7] Y. Aharonov and D. Bohm, Phys. Rev. **115** (1959) 485.
[8] Chan Hong-Mo and Tsou Sheung Tsao, *Some Elementary Gauge Theory Concepts*, World Scientific, Singapore 1993. A. Jaffe and C. Taubes, *Vortices and Monopoles*, Birkhauser, Boston 1980. For a recent review of vortices in high-temperature superconductivity, see G. Blatter et al., Rev. Mod. Phys. **66** (1994) 1125.
[9] C. Rebbi and G. Soliani, *Solitons and Particles*, World Scientific, Singapore 1984.
[10] N. Craigie et al. eds., *Monopoles in Quantum Field Theory*, World Scientific, Singapore 1982.
[11] S. Coleman, Uses of Instantons, in *Aspects of Symmetry*, Cambridge Univ. Press 1985.

CHAPTER 5

[1] N. Straumann, Deutschc Math. Verein Seminar on the History of Mathematics: H. Weyl's *Space-Time-Matter* (1992) (in German).

[2] C. N. Yang, in "Hermann Weyl's Contribution to Physics," in *Hermann Weyl*, ed. K. Chandrasekharan, Springer-Verlag 1980.

[3] See, for example, J. Bjorken and S. Drell, *Relativistic Quantum Mechanics I*, McGraw-Hill, New York 1964.

[4] R. Marshak, Riazuddin, and C. Ryan, *The Theory of Weak Interactions in Particle Physics*, Wiley-Interscience, New York 1969.

[5] R. Marshak and E. C. G. Sudarshan, Phys. Rev. **109** (1958) 1860. R. Feynman and M. Gell-Mann, Phys. Rev. **109** (1958) 193.

[6] I. Novikov and Y. Zeldovitch, *The Structure and Evolution of the Universe*, Chicago Univ. Press (1983). G. Gibbons, S. Hawking, and S. Siklos, *The Very Early Universe*, Cambridge Univ. Press, 1983. G. Steigman, *Cosmology Confronts Particle Physics*, Ann. Rev. Nucl. Sci. **29** (1979) 313.

[7] E. Majorana, Nuovo Cim. **14** (1937) 171.

[8] E. Wigner, Zeit. f. Physik. **53** (1929) 592.

[9] A. Einstein, Sitz. Ber. Preuss. Akad. Wiss (1928) 217.

[10] See, for example, N. Straumann, *General Relativity and Relativistic Astrophysics*, Springer-Verlag 1984.

CHAPTERS 6–10

[1] O. Klein, 1938 Conference on New Theories in Physics, held at Kasimierz, Poland 1938, reprinted in *1988 Conference on New Theories in Physics, Proc. XI Warsaw Symposium on Elementary Particle Physics*, eds. Z. Ajduk, S. Pokorski, and A. Trautman, World Scientific, Singapore 1989.

[2] Conference in Honour of H. A. Lorentz, Leiden, 1953. Proceedings in Physica, **19** (1953). A. Pais, p. 869.

[3] C. N. Yang, *Selected Papers with Commentary*, Freeman, New York 1983.

[4] C. N. Yang and R. L. Mills, Phys. Rev. **96** (1954) 191.

[5] C. N. Yang and R. L. Mills, Phys. Rev. **95** (1954) 631.

[6] C. N. Yang, "Hermann Weyl's Contribution to Physics," in *Hermann Weyl*, ed. K. Chandrasekharan, Springer-Verlag (1980).

[7] R. Shaw, Thesis, Cambridge University 1955.

[8] Ryoyu Utiyama, *Butsurigaku wa dokumade susundaka* (How Far Has Physics Progressed?), Iwanami Shoten, Tokyo 1983.

[9] Ryoyu Utiyama, *Ippan Gauge ba ron josetsu* (Introduction to the General Gauge Field Theory), Iwanami Shoten, Tokyo 1987.

CONCLUSION

[1] J. Bernstein, *Spontaneous Symmetry Breaking and All That*, Rev. Mod. Phys. **46** (1974) 7.

[2] J. Collins, *Renormalization*, Cambridge Univ. Press 1984.

[3] C. Quigg, *Gauge Theories of Strong, Weak and Electromagnetic Interactions*, Benjamin-Cummings, New York 1993.

[4] C. Rubbia, Rev. Mod. Phys. **57** (1985) 699.

[5] P. Langacker et al., Rev. Mod. Phys. **64** (1992) 87.

[6] Y. Nambu, *Quarks*, World Scientific, Singapore 1985. F. Halzen and A. Martin, *Quarks and Leptons*, Wiley, New York 1984.

[7] D. Politzer, Physics Reports **14C** (1992) 87.

[8] R. Taylor, H. Kendall, and J. Friedman, Rev. Mod. Phys. **63** (1991) 573, 597, 615.

[9] D. Denergi et al., Rev. Mod. Phys. **62** (1990) 1.

[10] M. Kobayashi and T. Masakawa, Prog. Theor. Phys. **49** (1973) 652.

Index

ABOUT THE AUTHOR

Lochlainn O'Raifeartaigh is Senior Professor at the Dublin Institute for Advanced Studies, where he teaches courses in Quantum Field Theory and Particle Physics. He is the author of *Group-Structure of Gauge Theory.*